I0065675

LAMPS & LIGHTING IN HAZARDOUS AREAS

Ian Staff

ELECTRICAL TRAINING CONSULTANT

First Edition published 2023

2QT Publishing

Stockport,

United Kingdom.

SK5 9BL

Copyright © Ian Staff 2022

The right of Ian Staff to be identified as the author of this work has been asserted by him in accordance with the Copyright, Designs and Patents Act 1988

All rights reserved. This book is sold subject to the condition that no part of this book is to be reproduced, in any shape or form. Or by way of trade, stored in a retrieval system or transmitted in any form or by any means, electronic, mechanical, photocopying, recording, be lent, re-sold, hired out or otherwise circulated in any form of binding or cover other than that in which it is published and without a similar condition, including this condition being imposed on the subsequent purchaser, without prior permission of the copyright holder

Cover design: Dale Rennard

Images supplied by author

Printed in the UK by Lightning Source UK Limited

ISBN 978-1-914083-66-2

About the Author

I am an Electrical Training Consultant and carry out Electrical Training for a company in Hull by the name of Humberside Offshore Training Association Ltd. (H.O.T.A.), Malmo Road, who, by the way, is one of my sponsors. Before my 15 or so years at H.O.T.A. as a Trainer /Assessor I was 38 years with BP, seven of those years as their Instrument/Electrical Training Officer in charge of all Instrument and Electrical Training and part of their Team in their Training Department where I obtained my Training and Assessing Qualifications.

The Maintenance Electrical Technician in Hazardous Areas (Full Set)

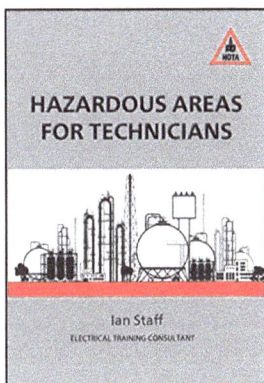

ISBN 978-1912014958   ISBN 978-1913071615   ISBN 978-1914083013   ISBN 978-1914083112

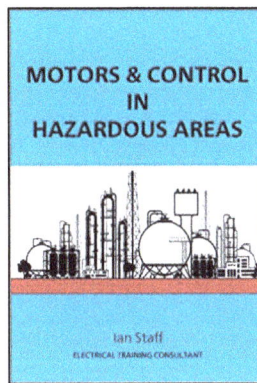

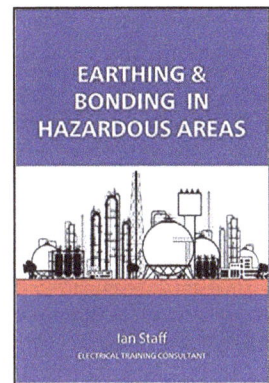

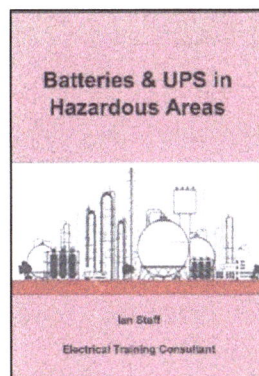

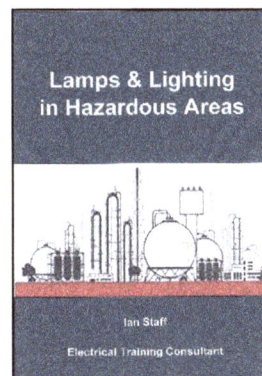

ISBN 978-1914083303   ISBN 978-1914083662

So far I have written these six books in this set showing the knowledge that, in my opinion, is required by an Electrical Technician on a Chemical Factory or a Platform. Having once been there I feel that these books, at a reasonable cost, written at Technician level would demonstrate the full spread of subjects dealt with in everyday life. The first book talks mainly about Atex, the IEC and EN/IEC Standards, Zones, Gas Groups, Temperature Classes etc. and sets the basis for the other five ie: Inspections of Electrical and Instrument Equipment, Motors and Control Circuits, Plant Earthing and Bonding System Types and Testing, Battery types and Maintenance including UPS Systems and finally Lamps and Lighting Systems shows all of the different Lamps/Light Bulbs available, where they are used and controlled in hazardous areas.

Introduction

This book: 'Lamps and Lighting in Hazardous Areas' is the sixth book in the series and explains the history of the electric lamp as well as diagrams and explanations of all lamps that are available.

My first book 'Hazardous Areas for Technicians' is very successful, explaining what Atex is and how Atex equipment is installed and maintained in Hazardous Areas as well as things like 'Cathodic Protection' etc.

My second book 'Inspections in Hazardous Areas' is mainly for Companies who carry out their own Inspections of Hazardous Area Electrical, Instrument and Mechanical Equipment and focuses how Electrical, Instrument and Mechanical Inspectors go about understanding standards, why the equipment is inspected and the terminology.

My third book Motors and Control in Hazardous Areas is all about the History of the electric motor and all of the different motors that Electrical Technicians may come across in their Hazardous Areas from DC to AC.

My fourth book 'Earthing and Bonding in Hazardous Areas' which like the others starts off with the history of the colours and symbols right up to how to carry out the installation and testing and why various tests are done. The book explains the importance of the Earthing and all of the different methods that can be installed into the ground to achieve this.

My fifth book 'Batteries and UPS in Hazardous Areas' deals will all of the different batteries that are available, how they work and where they are used. UPS systems that companies may have installed are also discussed.!

Content

Lamp History:

Different lamps or parts of lamps were developed by various Engineers and Physicists throughout history and we must be thankful to them for what we have in lighting today. Let us have a look at a few characters of history and their contributions, you may know more. When you are looking through just remember what year it was when they developed their equipment.

1710 – Francis Hauksbee (1616-1713): British Scientist conducted experiments with Static Electricity to produce a very low intensity light in a vacuum inside a glass globe.

1800 – Vasily V. Petrov (1761 – 1834): Russian Physicist developed a crude Carbon Arc Lamp.

1800 – Humphrey Davy (1778-1829): British Chemist and Inventor demonstrated a very basic Arc Lamp. He passed a current through a thin strip of platinum which like Tungsten had a very high melting point. The light was extremely intense but did not last long so was found not to be practical. It did demonstrate that light could be produced by electricity

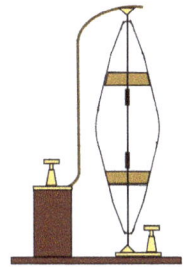

1806 – Davy redesigned and demonstrated a better lamp where the arc was struck between two Carbon Rods.

1835 – James Bowman Lindsay (1799-1862): Scottish Inventor developed an Incandescent Electric Lamp that could last quite a long time.

1835 – Charles Wheatstone (1802-1875): English Scientist tested Mercury Vapour Arcs.

1840 – Warren de la Rue (1815-1889): British Astronomer, Chemist and Inventor invented a lamp element. A coiled platinum wire which he encased in a vacuum in a glass tube.

1840 – Levett Ibbetson (1799-1869): English Inventor was the first person who used artificial light for photography called Limelight. (Where the saying originated 'in the Limelight')

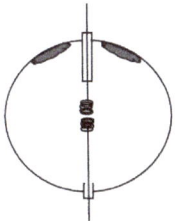

1841 – Frederick de Moleyns: British Inventor invented an Incandescent Lamp with a vacuum in a glass bulb with a Platinum Filament. In some circles this was credited as being the first practical incandescent lamp. One of the problems that he faced and even these days incandescent lamps face is a blackening of the inside top of the glass bulb. Because he used platinum as the filament this made his lamp very expensive. However this was considered a huge step forward!

1840 – Jean B. L. Foucault (1819 – 1868): French Physicist developed a mechanism for feeding Carbon Rod Electrodes in arc lamps.

1850 – Joseph Wilson Swan (1828-1914): British Physicist, Chemist and Inventor used carbonised paper filaments in a vacuum inside a glass bulb.

1856 – Heinrich Geissler (1814-1879): German Glassblower developed a glass tube which was the basis of the Fluorescent Tube.

1859 – Alexandre Edmond Becquerel (1820-1891): French Physicist developed the first Phosphor Coated Fluorescent Tube.

1860 – The measurement 'Candle power' was defined in the Metropolitan Gas Act 1860.

1860 – John Thomas Way (No details) developed a Mercury Vapour Lamp.

1865 – Hermann Sprengel (1834-1906): German Chemist invented a Mercury Vacuum Pump which other inventers used in their lamps to extend the life of the filament.

1869 – William Crookes (1832-1919): English Chemist experimented with a discharge tube with a partial vacuum putting 10,000 Volts along the tube. (Crookes Dark Space).

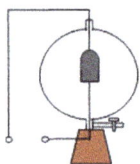 **1874** – Henry Woodward (1832-1921): Canadian Inventor together with partner Mathew Evans a Hotel Keeper, developed a light bulb consisting of a Carbon Pod connected by two wires in a glass tube, with a valve at the bottom, which they filled with Nitrogen. This paved the way for later inventions were developed and based on this one.

1874 – Alessandro Cruto (1847-1908): Italian Inventor developed the first Synthetic Filament.

1875 – Pavel NiKolayevich Yabloshkov (1847-1894): Russian Engineer and Inventor invented a lamp using two parallel carbon rods it was nicknamed 'Yablochkov Candle' His invention lit up the Avenue de l'Opera in Paris in this year.

1878 – William E. Sawyer (1850-1883): American Inventor and Albion Man invented a lamp with a Nitrogen filled bulb.

1878 – First home, belonging to Lord Armstrong, where electric lighting installed.

1878 – Joseph Wilson Swan (1828-1914): British Physicist, Chemist and Inventor developed a light bulb using treated cotton thread as the filament.

1879 – Charles F. Brush (1849-1929): American Inventor invented a Carbon Arc street light. The lights were first installed in Cleveland, Ohio and after their success were then installed in New York and later in London. Brush went on to found the Brush Electric Company. The Carbon Rods that produced the arc had to be adjusted from time to time as their tips burned and a limited electromagnetic adjuster was later incorporated. The design was later enhanced with copper plating.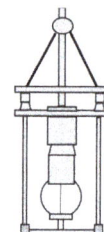

1879 – Thomas Edison (1847-1931): American Inventor designed a lamp with a very thin filament and a high resistance.

1879 – Niagara Falls lit up for the first time with Arc Lamps.

1879 – William Wallace redesigned the arc lamp Carbon Electrode to include other materials.

1880 – Thomas Edison (1847-1931): American Inventor designed a lamp with a Bamboo Filament that lasted over 1000 hours.

1881 – Thomas Edison (1847-1931): American Inventor patented the Edison Screw Cap.

1882 – Edwin Scribner (no information) patented a Chlorine filled lamp which was not successful. Possibly the first time Halogens were attempted in lighting.

1892 – Lee Arons (1860-1919): German Physicist developed an experimental Mercury Vapour Lamp.

1896 – H. J. Dowsing (1859-1931) & H. S. Keating patented what has come to be known as the first true Mercury Vapour Lamp.

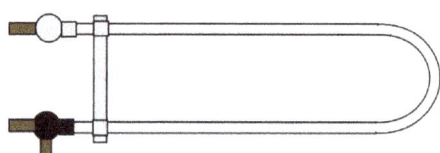 **1901** – Peter Cooper Hewitt (1861-1921): American Electrical Engineer and Inventor patented a Mercury Vapour Lamp but the lamp was not a success as the colour was poor and large ballasts were needed to control the current.

1903 – Robert W. Wood (1868-1955) American Physicist developed the Wood's Black Lamp.

1906 – Kuch & Retschinsky (no information): Developed a high-pressure Mercury Vapour Lamp.

1910 – William David Coolidge (1873-1975): American Physicist and Engineer developed a long lasting Tungsten Filament whilst working for the General Electric Company.

1911 – Georges Claude (1870-1960): French Engineer and Inventor invented the Neon Lamp.

1912 – Charles Proteus Steinmetz (1865-1923): German Electrical Engineer invented metal Halide Lamps.

1915 – Irving Langmuir (1881-1957) :American Physicist, Chemist and Engineer, whilst working for the General Electric Company, invented an improvement of the Incandescent Lamp with a Tungsten Filament in a gas filled glass bulb.

1915 – Elmer Ambrose Sperry (1860-1930): American Inventor developed a Carbon Arc Spotlight of many thousands of candle power used in naval applications.

1917 – Albert Einstein (1879-1955): German Physicist first brought forward the theory of the possibility of the Laser.

1920 – Arthur H. Compton (1892-1962): American Physicist invented the Low Pressure Sodium Lamp. A spherical chamber with Sodium Metal. When the electrodes heated up the lamp the Sodium Metal would vaporise and the lamp would have a yellow glow. Sodium is very corrosive and there were problems with the seals.

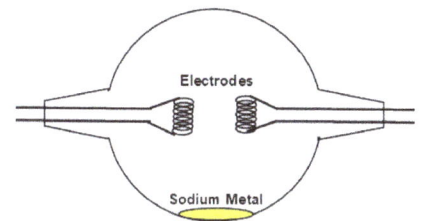

1921 – Marvin Pipkin (1889-1977): American Chemist and Engineer developed a way to frost the inside of the glass of a light bulb, but there were flaws in his process.

1927 – Edmund Germer (1901-1987): German Inventor patented a Fluorescent Lamp. He was assisted by Friedrich Meyer and Hans J. Spanner.

1931 – Marcello Pirani (1880-1968): German Physicist working for Osram developed a Sodium resistant glass.

1931 – Dr. Albert W. Hull (1880-1966): American Physicist obtained a patent for a low pressure Discharge Lamp paving the way for the Fluorescent Lamp.

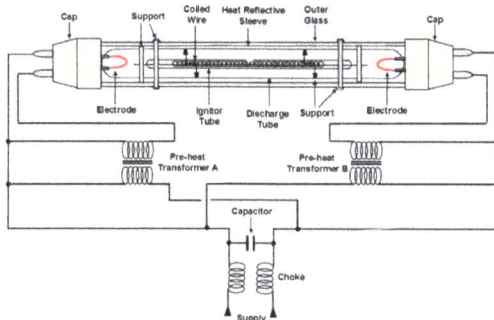

1932 – Developed in Germany, the diagram left is an example of the first self-igniting low pressure Sodium Lamps. Not that dissimilar in circuitry from the modern bi-pin fluorescent lights. The discharge tube is elongated so there is no 'U' tube here that came later. Once the power is put onto the fitting from the choke to the electrode pre-heat transformers heating up the electrodes and causing an ionisation of the gas in the immediate vicinity.

1940 – Fluorescent Tube presented at World Fair.

1955 – Robert L. Coble (1928-1992): American Ceramic Scientist developed a material called 'Lucalox' (Aluminium Oxide Ceramic)

1959 – Elmer Fridrich (1920-) & Emmet Wiley (1927-) patented a Halogen Lamp.

1960 – Fredrick Moby: American GEC Electrical Engineer patented an improved Tungsten Halogen Lamp.

1961 – Joel S. Spira (1927-2015): American inventor invented a lighting dimmer based on dropping the voltage. Now we can dim lighting using several methods including varying waveforms.

1961 – James Biard (1931-): American Electrical Engineer and Gary Pittman invented the very first Light **E**mitting **D**iode.

1962 – Professor Nick Holonyack (1928 -): American Inventor invented the first basic LED.

1964 – W. Louden, K. Schmidt & E. Homonnay GEC Scientists invent the first High Pressure Sodium Lamp. (HPS).

1964 – GEC put onto the market the first High Pressure Sodium Lamp.

1976 – Edward E. Hammer (1931-2012): American Engineer bent a fluorescent tube to represent a spiral opening the door to compact Fluorescent Lighting and so inventing the very first Compact Fluorescent.

1980s – Metal Halide Lamp invented.

1985 – GEC Scientists improved the High Pressure Sodium (HPS) efficiency.

1990s – Compact Fluorescent Lamp invented.

1991 – Shuji Nakamura (1954 -) American Electronic Engineer & Isamu Acosoci Amano (1929-2021) Japanese Engineer & Physicist developed the blue LED.

1991 – Michael Ury & Charles Wood both Engineers developed Sulphur Lamp technology.

1993 – Shuji Nakamura (1954 -) American Electronic Engineer after much development he created the commercial LED lamp.

2006 – Much brighter LEDs were developed.

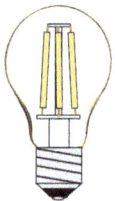

Atomic Structures:

Before we can fully understand the workings of our lamps/bulbs we need to know a little about the atomic structure of materials, so that the insulation, conduction and semi-conduction of metals and gases used can be better understood and why inventors used them in the first place. I hope that I have not made the following explanations too complex.

Molecules:

A '**molecule**' is a small unit of an '**Element**' or '**Compound**'. Simply, we can say that a molecule is a group of two or more '**atoms**'. The structure of a molecule is held together by what is called '**Covalent Bonding**'. So for instance if we take water which has a chemical formula of H_2O what this is telling us that it is a molecule made up of 3 atoms, two of Hydrogen (H_2) and one of Oxygen (O). Carbon Dioxide CO_2 is one Carbon Atom (C) bonded to two Oxygen Atoms (O_2)

Another formula might include **$2H_2SO_4$** (2 molecules of Sulphuric Acid) so what would be the meaning here? Well the first large 2 means that there are 2 molecules. (Remember, molecules contain 2 or more atoms) and the next figures and smaller numbers show there are 7 atoms. In this case, as mentioned, the chemical symbol is for 2 molecules of Sulphuric Acid (**$2H_2SO_4$**) containing 2 atoms of Hydrogen (H_2), 1 atom of Sulphur (S) and 4 atoms of Oxygen (O_4)

Atoms:

So where does the above leave us with atoms? Well we can say as a definition that an atom is the smallest unit of an element or compound. Atoms are composed of a nucleus which has Protons and Neutrons inside and Electrons in orbit around the nucleus. The number of Electrons will always be equal to the number of Protons and this will be their '**Atomic Number**' on the **Periodic Table of Elements** with some examples below.

1	19	29	13
H	**K**	**Cu**	**Al**
Hydrogen	Potassium	Copper	Aluminium
Non Metal	Alkali Metal	Transition Metal	Post Transition

The letters on the Periodic Table left do not always seem to denote the Element. It tends to pick out the older names for Elements i.e.: 'K' for Potassium comes from Kalium which is Latin for Potash.

Looking at the diagram above of sections from the Periodic Table we see Hydrogen (H) has 1 Proton and 1 Electron, it is a **non-metal.** Potassium (K) has 19 Protons and 19 Electrons and is an '**Alkali**' Metal. Copper (Cu) has 29 Protons and 29 Electrons and is a '**Transition**' Metal. Transition meaning that they are in the centre (groups 3–12) of the Periodic Table, have Catalytic Properties, high Melting Points and Densities. Aluminium has13 Protons and 13 Electrons and is Post Transition meaning it is in groups 13-15 on the Periodic Table.

Protons have a positive (+) charge, Neutrons have a neutral charge, both of these never leave the nucleus. **(Just out of interest Protons and Neutrons are made up of sub-atomic particles called Quarks!)** Electrons however in orbit around the nucleus will have a negative (-) charge.

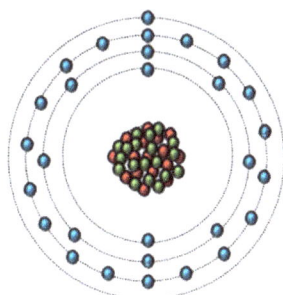

On the left is a Copper (Latin **Cu**prum Cu) Atom - Atomic Number 29. The number of Protons is always equal to the number of Electrons so our Copper Atom has 29 Protons so it will have 29 Electrons arranged in orbits and in this case there are 4 orbits. The orbits are called '**Shells**' and are arranged so the closest to the Nucleus will have up to 2 Electrons, the second orbit several Electrons other orbits can take many more and will make up the total number of Electrons for that element. The Electrons that are the easiest to move are the '**loosest**' and are the ones in the outer orbits or '**Valance**'. Protons and Neutrons remain locked in the Nucleus of the atom and do not move.

Sum up Questions on Incandescent Lamps:

Q1 – Are Electrons positively charged, negatively charged or neutral?

A1 – NEGATIVELY CHARGED. They are in orbits called 'Shells' around the nucleus. Protons are positively charged and Neutrons as the name suggests are neutral charge.

Q2 – Can you get smaller particles than Electrons, Protons & Neutrons?

A2 – YES. Most books might say no, but the three mentioned particles are made up of what are called **'Quarks'**

Q3 – How are atoms donated in a formula?

A3 – ATOMS ARE SHOWN AS SMALL NUMBERS OR NO NUMBERS. So if we take a common formula that everyone knows of water which is H_2O this is telling us that water is made up of 3 atoms in all; 2 atoms of Hydrogen (H_2) and 1 atom of Oxygen (O). Another one is Carbon Dioxide CO_2 which is again 3 atoms - 1 atom of Carbon (C) and 2 atoms of Oxygen (O_2).

Q4 – What is a Molecule?

A4 – A SMALL UNIT OF AN ELEMENT OR COMPOUND. When you see a formula which is shown as **$2H_2SO_4$,** this is Sulphuric Acid. This compound has 2 molecules with 7 atoms, 2 atoms of Hydrogen, 1 atom of Sulphur and 4 atoms of Oxygen.

Q5 – What holds the structure of an atom together?

A5 – COVALENT BONDING.

Q6 – When an Element has an 'Atomic Number' what does that mean?

A6 – THE NUMBER OF ELECTRONS & PROTONS. The number of Electrons in the shell orbits will always equal the number of Protons in the nucleus.

Q7 – How do we determine how many Shells the atom has got and how many Electrons in each shell?

A7 – BY THE FORMULA $2(n)^2$ This is a very complicated answer. In the formula 'n' is the shell number so in the first shell it would read $2(1)^2 = 2$. So there would be 2 Electrons in the first shell. In the second shell it would read $2(2)^2 = 8$ so there would be 8 Electrons in the second shell and so on. That is as far as we will go here.

Q8 – What is an atom's outer Electron orbit called?

A8 – THE VALANCE.

Q9 – What are 'Loose' Electrons?

A9 – THE ONES IN THE OUTER SHELL OR VALANCE. When a current is flowing it is these Electrons that move the easiest.

Q10 – Why do Element Letters on the Periodic Table sometimes not match the substance? For Example Copper (CU) Potassium (K) Hg Mercury etc.

A10 – OBSOLETE NAMES & LATIN NAMES. For instance Copper is **Cu** because **Cuprum** is an obsolete name for Copper. Potassium **K** because **Kalium** is Latin for Potash. Mercury is **Hg** because of the Latin & Greek name **Hydrargyrum.** Hydrargyr(us) Hydrargyr(um)

What is a Conductor?

Again before we go too much further, what makes a '**Conductor**' different from an '**Insulator**'? Let us firstly have a look at two popular metals that are very good conductors and are used extensively in the production of electric cables and current carrying equipment.

Question: What makes these **Electrical Conductors** better than each other?

Answer: Their Atomic Structure.

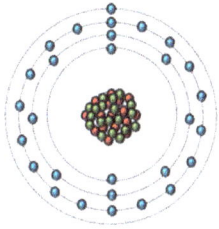

Copper
Chemical Symbol Cu
Atomic Number 29
Nucleus 29 Protons
Nucleus 34 Neutrons
4 Orbits of Electrons
29 Electrons

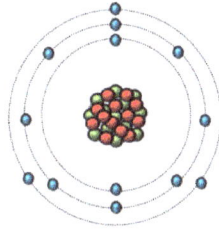

Aluminium
Chemical Symbol Al
Atomic Number 13
Nucleus 13 Protons
Nucleus 14 Neutrons
3 Orbits of Electrons
13 Electrons

The **Atomic Number** on the **Periodic Table** i.e. Copper 29 and Aluminium 13, refers to the number of Protons in the nucleus.

The number of Protons as seen in the diagrams on the left, is always equal to the number of Electrons. So our Copper atom has 29 Protons and therefore will have 29 Electrons in 4 orbits. Aluminium has 13 Protons so will have 13 Electrons arranged in 3 orbits and so on.

The orbits are called '**Shells**' and are arranged so the closest to the nucleus, the 1st shell, will have up to 2 Electrons, the second orbit, 2nd shell several electrons other orbits can take many more and will make up the total number of Electrons for that Element.

So if we look at the atomic properties of these metals, we find that the metalic bonds of the Electrons are very 'loose' and if a voltage is applied from, say, a battery, they will freely move around the material from atom to atom, hence we have now got an electric current flowing. The Electrons that are the easiest to move of course and are the 'loosest' are the ones in the outer orbits or '**Valance**' which usually do not number many. The more loose Electrons in the outer or 'Valance' orbit the easier they move. Protons and Neutrons remain locked in the nucleus of the atom and do not move in the event of applying a voltage. When Electrons move they leave '**holes**' for others to move into.

How do we determine how many shells an Atom has got? By the formula $2(n)^2$. This is a complicated answer. In the formula, 'n' is the shell number so in the first shell it would read $2(1)^2 = 2$ so there would be 2 Electrons in the first shell. In the second shell it would read $2(2)^2 = 8$ so there would be 8 Electrons in the second shell and so on. That is as far as we will go here.

The above metals or conductors, are used extensively for electric cables in the electrical world. '**Copper**' is used in general cabling more than '**Aluminium**', which although is not used much in wiring equipment in domestic and industrial premises, but because it is much lighter than copper, is used for high voltage from pylon to pylon on the national grid.

Metal:	Chemical Symbol:	Atomic Number:	Protons:	Neutrons:	Orbits:	Electrons:
Silver	Ag	47	47	60-62	5	47
Gold	Au	79	79	118	6	79
Zinc	Zn	30	30	35	4	30
Nickel	Ni	28	28	31	4	28
Brass	Alloy of Copper and Zinc					
Bronze	Alloy of Copper and mainly Tin					
Iron	Fe	26	26	30	4	26

The table above shows some other metals, besides Copper and Aluminium, which conduct electricity best at the top to fair at the bottom! We do not make our cable conductors of Gold or Silver for obvious reasons although certain contacts of more expensive relays/contactors may have a Silver coating.

Sum up Questions on Conductors:

Q1 – What makes the Electrons move to make a current in a conductor?

A1 – **A VOLTAGE.** By applying a voltage, AC or DC, to a conductor will result in an electric current flow.

Q2 – Why do Conductors actually work?

A2 – **LOOSE ELECTRONS.** If we look at the atomic structure on the previous page and the Electrons in orbit around the atom nucleus, the Electrons in the outer orbit, or the valance, are easily moved from atom to atom when a voltage is applied. They are called 'Loose' Electrons as their bonds to the atom are not tight as they would be in an Insulator.

Q3 – What materials are the best conductors?

A3 – **METALS, CARBON ETC.** Most metals will conduct electricity some better than others. Copper and Aluminium you may recognise in the electrical world as good conductors. It may be that Gold and Silver are even better, but expensive. Other materials will also conduct, for example, Carbon which is used for motor & generator brushes etc.

Q4 – What makes one conductor different from another?

A4 – **THEIR ATOMIC STRUCTURE.** If we take Copper and Aluminium as an example. Copper's Atomic Number on the Periodic Table is 29 which means that it has 29 Electrons in orbits and 29 Protons in the nucleus.

Aluminium's Atomic Number is 13 which means it has 13 Electrons in orbits and 13 Protons in the nucleus. (see previous section)

Q5 – Is water a good conductor of electricity?

A5 – **NO.** It is not exactly a bad conductor but although it will conduct it does spread all over which makes it dangerous. If it was a good conductor we could have hosepipes full of water instead of cables which would be much cheaper. Distilled water is of course an Insulator.

Q6 – Do conductors do their job better on AC or DC?

A6 – **EITHER.** DC is the most efficient voltage, but its problem is a phenomenon called 'Volt Drop' over long distances. Hence the National Grid is 450,000 Volts AC. Just out of interest they have just run a cable 450 miles from Norway to Northumberland to power 1.4 million homes, it will have to be around 500,000 Volts AC. If it was DC we probably could not be able to run a train set on this end.

Q7 – Are the conducting cables on the National Grid Copper?

A7 – **NO.** Aluminium cables are used to go from pylon to pylon because Aluminium is less dense and so much lighter. Copper cables would be far too heavy.

Q8 – Are Industrial and domestic power and lighting cable conductors Copper?

A8 – **YES.** For industrial and domestic conducting circuits, the wiring for lighting and most equipment is Copper. Aluminium cable cores cannot be used under 16mm² in a hazardous zoned area.

What is a Semi-conductor:

A Semi-conductor is in actual fact an 'Insulator' until certain conditions change and it becomes a 'Conductor'. That could be a change in voltage, temperature, light or the addition of impurities called doping (such as Arsenic or Phosphorus). Conductors have the ability to have current flowing through them by having 'free' electrons in an outer orbit, so by doping, we add an element that has 'free' electrons to our compound with no 'free' electrons to encourage it to conduct i.e.: the free electrons from the doping will move and hence provide an electric current.

Arsenic & Phosphorous would give 'N' type doping as the free electrons would be a Negative charge. Boron for instance would be a 'P' type doping, this encourages **'holes'** in the **Covalent Bond** (Below) which would leave the electrons floating around with nothing to bond to so would give a positive charge

Zener Diodes are a good example of a Semi-conductor that changes with voltage. The Zener Barrier that contain the Zener Diodes protects the Intrinsically Safe equipment in the zoned area from a dangerous rise in voltage. One lamp of course that uses this technology is the **LED Lamp.** Integrated Circuits and Photovoltaic Cells (Solar Cells) are another abundant use of semi-conductors.

Two Semi-conductors that readily spring to mind are Silicon and Germanium, but these emit their energy in heat and not light. Other semi-conductors which emit their energy in the form of light are used in LED lamps and these are Gallium Arsenide (GaAs), Gallium Phosphide (GaP) and Gallium Arsenide Phosphide (GaAsP) depending upon what colour LED lamp is required.

Silicon (Si):

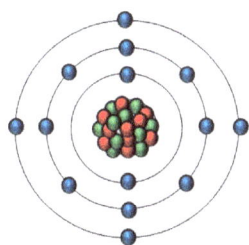

Let us first look at Silicon. This compound as you can see from the atomic diagram on the left has 14 electrons in 3 shells which means that it will have 14 Protons in its nucleus. Silicon has an operating temperature of above 100 degrees centigrade which makes it attractive for many uses where temperature is a problem. The electrons in the orbit shells are in what is called Covalent Bonds, with neighbouring electrons in their atoms. So unlike metals they are not free to just move around and form a sort of lattice.

Germanium (Ge):

Now let us look at Germanium. This compound, as you can see from the atomic diagram on the right, has 32 electrons in 4 shells which means that it will have 32 Protons in its nucleus. Germanium has an operating temperature of above 70 degrees centigrade which although is lower than Silicon, makes it attractive for many uses where temperature is a problem. The electrons in the orbit shells are in what is called Covalent Bonds with neighbouring electrons in their atoms. So unlike metals they are not free to just move around and form like a lattice.

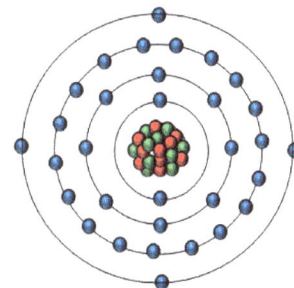

So a Semi-conductor is a compound insulator usually of Silicon or Germanium whose atoms do not have 'free' electrons. Making it conduct can be done - **a)** by 'Doping' i.e. adding Arsenic or Phosphorous to add free electrons and create an 'N' type (Negative) or something like Boron to create 'holes' in the Covalent Bonding which would be a 'P' type (Positive). Doping the Semiconductor makes them **'Extrinsic'** and the other means of electron movement is called **'Intrinsic'** as the next actions are called; **b)** where a breakdown voltage is inserted to make the Electrons move out of their bonding or **c)** introducing light to cause enough EMF for Electron movement.

Without this technology there would be no electronic items such as Transistors, Diodes or in fact any Integrated Circuits so Semi-conductors play a very important part in our electronic technology.

Sum up Questions on Semi-conductors:

Q1 – Is a Semi-conductor an insulator or conductor?

A1 – **AN INSULATOR:** Until something else is done to make it a conductor. I.e. doping, Breakdown Voltages etc.

Q2 – If the doping was Arsenic or Phosphorous what type would this doping be?

A2 – **'N' TYPE:** To install **'free'** electrons into the Semi-conductor would give it a negative charge.

Q3 – If the doping was Boron what type would this doping be?

A3 – **'P' TYPE:** To install 'holes' in the **'lattice'** would give it a positive charge.

Q4 – What are the most common Semi-conductors?

A4 – **SILICON AND GERMANIUM:** Which I am sure you may have heard of.

Q5 – Are Semi-conductors used in lighting?

A5 – **YES:** They are used in LED lamps. The Gallium range of Semi-conductors, **NOT** Silicon or Germanium in this case.

Q6 – What are 'Extrinsic' Semi-conductors?

A6 – **DOPED:** These are called **'Extrinsic'** whilst others like excess voltage or light are called **'Intrinsic'** as below.

Q7 – What are **'Intrinsic'** Semi-conductors?

A7 – **OTHER MEANS THAN DOPED:** Which might be changed by voltage, light etc.

Q8 – Can Semi-conductors operate in high temperatures?

A8 – **YES:** Silicon can operate at the temperature over 100ºC and Germanium at around 70ºC.

Q9 – What do they call the Atomic 'Bonding' of Semi-conductors?

A9 – **COVELANT BONDING:** A sort of **'lattice'** where there are no 'Free' electrons until the Semi-conductor is doped.

Q10 – What does Silicon's Atomic Number 14 mean?

A10 – **14 ELECTRONS:** In the shells around the nucleus there are 14 Protons that remain in the nucleus. This goes for any Element or Compound on the Periodic Table.

Q11 – Can the Semi-conductor be made to turn on and off?

A11 – **YES:** The voltage and light operated Semi-conductors will do this all of the time.

Q12 – Where are Semi-conductors most used?

A12 – **INTEGRATED CIRCUITS:** Also used in most electronic components i.e. Transistors, Diodes etc. and in Intrinsically Safe protection such as Zener Barriers.

What is an Insulator?

We looked previously at popular conductors of electricity. Now let us look at '**Insulators**'.

Question:- What makes one material an 'Insulator' and another a 'Conductor'?

Answer:- Their Atomic Structure.

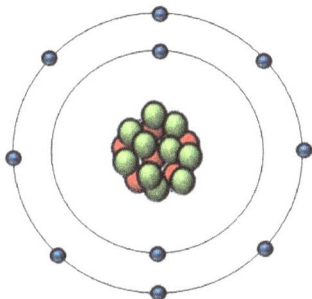

The number of Protons as you will see in the diagram on the left, is always equal to the number of Electrons. So our particular insulator atom has 10 Protons and so it will have 10 Electrons and in this case we have 2 orbits or shells.

The orbits are called '**Shells**' and are arranged so the closest to the nucleus in our case has up to 2 Electrons, the outer orbit or in our case 'Valance' has 8 Electrons. Different materials will have different numbers of orbits or shells and a different number of Electrons and Protons.

So if we now look at the Atomic properties of these insulators we find that the bonds of the electrons are very '**tight knit**' and if a voltage is applied from, say, a battery there are no 'loose' electrons so they will **NOT** freely move around the material from atom to atom, so there will be no electric current flowing.

In conductors, which are usually metals as mentioned, the electrons that are the easiest to move of course are the 'loosest' and are the ones in the outer orbits or valance which usually do not number many. **Here, the Electrons in the Valance do not move.** Protons and Neutrons, as in the case of the conductor, remain locked in the nucleus of the atom and do not move at all.

Several electric cable insulations are listed in the table to the right. Some you may recognise, like PVC which is the most common cable insulation, as the grey twin and earth in houses would be of this insulation. Sometimes the outer sheath may be a different material to that which covers the conductors.

Initials:	Full Name:
PVC	Polyvinyl Chloride
EPL	Ethylene Propylene Rubber
XLPE	Cross Linked Polyethylene
CPE	Chlorinated Polyethylene
PTFE	Polytetrafluoroethylene

Please remember that insulation can break down for many reasons i.e. dampness & water. Water is **NOT** a very good conductor of electricity otherwise we could have hosepipes full of water for carrying our electricity which would be much cheaper instead of Copper. The fact is that water **WILL** conduct and speads all over which makes it dangerous. High voltage can also break down an insultation if high enough.

Paper	Wax
Wood	Tufnell
Nylon	Paper
Plastic	Distilled Water
Glass	Porcelain
Teflon	Oil

On the left are several more insulators, many of which you will recognise and most of these are used in the world of electricity in either equipment, batteries and cables.

As mentioned above the solid Insulators on the left only do their job correctly if they are **DRY**. Get them damp and they could become conductors!

Who discovered Electrons, Protons & Neutrons:

1 – Electron: Negative Charge, Discovered in 1897 by J. J. Thompson

2 – Proton: Positive Charge. Theorised in 1815 by William Prout.

3 – Neutron: No Charge. Discovered in 1932 by James Chadwick.

Sum up Questions on Insulators:

Q1 – What makes the Electrons in an insulator different to a conductor?

A1 – **THE BONDING:** The bonds of the Electrons are very **'tight knit'** and if a voltage is applied from, say, a battery, they will **NOT** freely move around the material from atom to atom, so there will be no electric current flowing. So no 'free' Electrons in a valance.

High voltage of course will always break down an insulator. If we take an example of **SOME** electrical screwdrivers, there is a warning not to be used on live systems over a certain voltage.

Sometimes this may be an end to end flashover of the high voltage arc rather than the insulator actually breaking down.

Q2 – What materials are the best Insulators?

A2 – **NON CONDUCTORS:** PVC, plastic, rubber, wood, paper, glass, nylon, plastic, Teflon, distilled water, porcelain, vitrite & oil. Some of these materials are used in the manufacture lamps, lamp holders and wiring. (Glass, vitrite, porcelain, plastic)

All of these examples will become conductors if they become damp or the voltage gets too high.

Q3 – What makes one Insulator different from another?

A3 – **THEIR ATOMIC STRUCTURE:** The number of Electrons, Protons and Neutrons. Have a look at insulators on the Periodic Table.

Q4 – Do insulators do their job better on AC or DC?

A4 – **EITHER:** Both AC & DC equipment require insulation of one kind or another if we take electric wiring the insulation on the outside of the wire is the same be it AC or DC.

Q5 – Why is distilled and deionised water an insulator?

A5 – **NO IMPURITIES OR CONTAMINANTS:** It is the impurities in water such as salts that makes it a conductor although not a very good one. By adding salt will dramatically make the water a conductor so just think about sea water.

Q6 – Is air a good insulator?

A6 – **YES:** Otherwise switches, circuit breakers and many other electric/electronic devices would not work. I must stress that the air must be **DRY!**

Q7 – What is dielectric strength?

A7 – **BREAKDOWN VOLTAGE:** Put simply the highest voltage known to breakdown a particular insulating material.

Q8 – What is the best insulator ever?

A8 – **POSSIBLY A PERFECT VACUUM:** This contains no impurities at all. Its dielectric strength will be enormous, hence Vacuum Contactors and Circuit Breakers.

Q9 – If air is an insulator and voltage can jump how far will 240 volts **RMS** jump?

A9 – **IT WILL NOT:** At this low voltage you would actually have to touch it to get a shock.

What is Electron Flow?

Electron flow occurs in two entirely different types of electricity, namely AC (**A**lternating **C**urrent) which comes out of say a 13Amp socket in your house and DC (**D**irect **C**urrent) which is a battery voltage. DC is the most efficient of the two.

So we know that Electrons move when a voltage is put onto a conductor, so it is these particles that interest us most. Electrons have a negative (-) charge, Protons have a positive (+) charge and Neutrons have no charge. Electrons move and Protons and Neutrons remain stable in the Nucleus of the atom.

Let us look at a typical copper conductor:

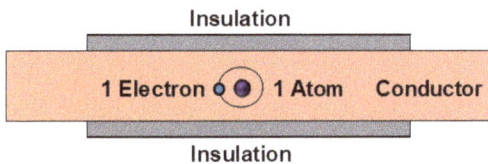

Left is a conductor which let us say, is the Copper conductor of an electric cable. The grey is the insulation. I have for, ease of explanation, only drawn 1 atom with 1 Electron. Let us put a DC voltage onto the conductor and see what happens.

Looking at the diagram on the right, when putting a DC voltage onto the conductor you can see that the Electron, being negatively charged, has now moved from the atom towards the positive side of the conductor. Remember, the Copper conductor is made up of billions of atoms **ALL** with '**Free**' Electrons, so they will **ALL** move in the same direction when the voltage is applied. We now have Electron flow or current flow which is described below:

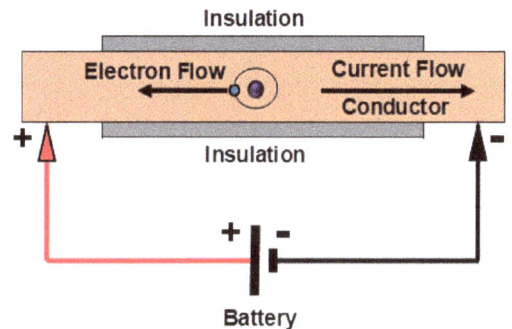

I am now going to complicate things a little! **Experiments** undertaken by **Benjamin Franklin** indicated that the current flow is in the opposite direction to the Electron flow. Now remember this is based on a theory by a great man but in my opinion '**it really does not matter**' How much current flows depends upon the voltage and what is called the 'resistance', which are measured in Ohms (Ω), of our conductor.

We mentioned earlier - amps (A), voltage (V), resistance/Ohms (Ω) and electrons:

Amps:

Current flow is measured in amps (A). The amp, short for **Ampere**, is named after **André-Marie Ampère** who was a French Physicist and Mathematician. If we looked at an equivalent water system then we could actually, for our analogy, take the water flow as the amperage.

Volts:

Volts are named after **Alessandro Volta** an Italian Physicist. So in our analogy, the pressure behind the water and the force that is pushing it through the piping would be the voltage.

Ohms:

The resistance of the conductor would be measured in Ohms (Ω) named after **Georg Simon Ohm** who was a German Physicist. Again in our analogy we could say that any restrictions in the pipework i.e. the pipe narrowing or going through a filter etc. would be our resistance.

Electron:

Discovered in 1879 by J. J. Thompson, when working on a Cathode Ray tube.

Sum up Questions on Electron Flow:

Q1 – Is it volts or amps that actually flow?

A1 – **CURRENT (AMPS):** Flows Positive to Negative.

Q2 – Is Current flow and Electron flow the same thing?

A2 – **NO:** According to Franklin. Electron flow is the opposite to the Current flow. Electron flow is from Negative to Positive!

Q3 – What initiates current flow?

A3 – **VOLTAGE:** If we take an analogy of water then the water flow is the current and the water pressure is the voltage.

Q4 – What Initiates Electron flow?

A4 – **EMF (VOLTAGE):** Electromotive Force.

Q5 – Are Electrons Positive or Negative charged?

A5 – **NEGATIVE:** Electrons have a Negative charge, Protons have a Positive charge and Neutrons have a Neutral charge.

Q6 – When was the Electron discovered and who by?

A6 – **J. J. THOMPSON IN 1879:** When he was working on a Cathode Ray tube.

Q7 – Do we get Electron flow in AC or DC?

A7 – **BOTH:** But in DC they only travel in one direction.

Q8 – Can Electrons flow without a voltage?

A8 – **NO:** It is obviously possible to have a voltage without electron or current flow, but it is not possible in a standard circuit to have Electron or current flow without a voltage!

Although I have actually stated '**NO',** in theory 'Free' Electrons that are in the outer valance are constantly on the move as they circle their atoms and maybe even filling random holes from atom to atom, but by putting a power source onto them they all move in the same direction along with the current.

Q9 – What are 'Excited State' Electrons?

A9 – **ELECTRONS THAT JUMP ORBITS:** Which they do from inner orbits when they absorb energy.

Q10 – Can magnetism cause electrons to flow?

A10 – **YES:** Motors and generators use magnetic fields to work. By moving a magnet in and out of a coil can cause an EMF and this would be called 'Electromagnetic Induction'.

Emil Lenz conducted some experiments in this subject. Induction lamps also use magnetism around their Induction Coils.

Q11 – Have Electrons got any mass?

A11 – **VERY LITTLE:** Which is why they move easily when energy is applied.

AC Voltage:

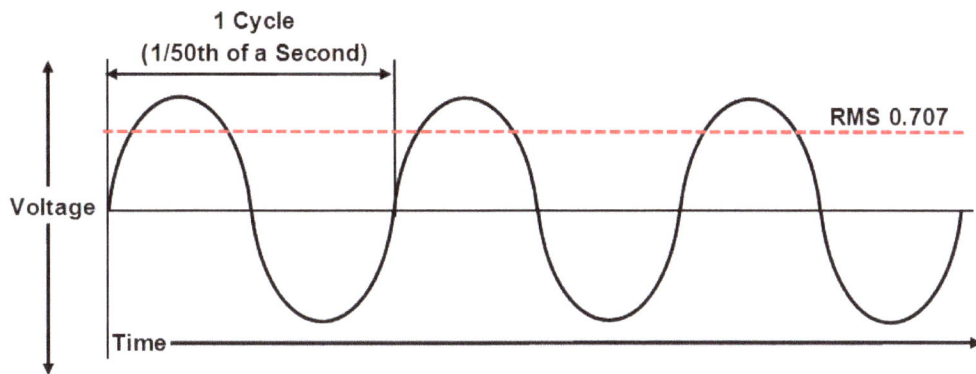

The diagram above shows a number of cycles in the UK and the frequency of the electricity shows 50 Cycles per Second (CpS) or 50 Hertz (Hz) as we call it now. In the USA this would be 60Hz. So each cycle of AC electricity would go up to positive peak and down through zero to negative peak which we know as a sine wave. AC voltages can easily be changed using devices called transformers, something that would be very difficult with DC as transformers only work with AC.

Let us look at DC for a moment. Here there is just a positive (+) and a negative (-) no matter how large or small the voltage is. With AC it is quite different as there are 3 phases, each 120 degrees out of phase with the other two. The colours used to be red, yellow and blue, but now they are brown, black and grey. Many industrial items of equipment are three phase, including the majority of electric motors. Domestic houses use one of these phases, which is called single phase, and the wiring is live (brown) and neutral (blue).

AC voltage is used in our homes and is transmitted in very high form, around 450,000 volts, all over the country by the national grid and transformed down through substations. AC is used by the grid as a voltage that can be transmitted many miles without much volt drop. They have just opened the longest cable run in history around 450 miles from Norway, where they use hydro (water) power, to Northumberland in the UK it will be AC and will power 1.4 Million homes.. If it was DC, with the volt drop I do not know what we would be left with by the time it got to the UK.

I have put a dotted red line on the diagram above which is the **Root Mean Squared Value or 0.707.** This is the most efficient or effective part of the sine wave or the DC equivalent section. When test instruments are put onto a circuit they read this **RMS** value not the peak value. Even electricity meters read this value. This could also be called the DC equivalent voltage of an AC Sine Wave.

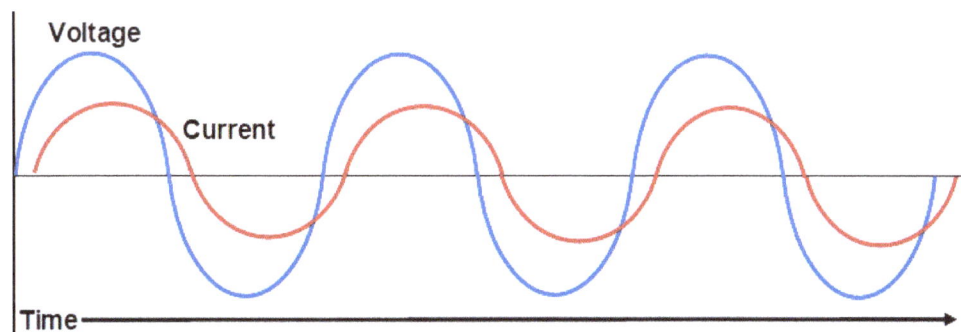

I often get asked if the sine wave is voltage what has happened to the current? Well looking at the diagram above, the blue sine wave is voltage and red sine wave is current. You will notice that the red sine wave is lagging the voltage which will tell us that this is an 'inductive' load such as a coil. If we got a capacitive load then the current would lead the voltage. The cosine of the angle by which the current lags or leads the voltage is what is called the **'Power Factor'** which suppliers try to keep to as near **'Unity' (1)** as possible. Usually industrial power factors run at around 0.8.

Sum up Questions on AC Voltage:

Q1 – What is the frequency of an AC Voltage?

A1 – 50 HERTZ: In this country the frequency, or Cycles per Second as it used to be called, is now 50Hz. We are talking about one cycle being 1/50th of a second. It is demonstrated on any drawings as a Sine Wave, but the waves can be all sorts of shapes from triangular to square which may be what an inverter produces before other equipment is used to make them roundish. In the USA this could be 60Hz.

Q2 – Why do we use AC?

A2 – EASIER TO TRANSMIT & TRANSFORM: AC can be transmitted over very long distances and can be transformed up and down quite easily. Transformers will not work on DC, so controlling it would be extremely difficult and any distance would cause volt drop. They have just run a cable approximately 450 miles from Norway to Northumberland to power 1.4 million homes, it would have to be around 500,000 Volts AC otherwise if it was DC we could probably not be able to run a train set on this end.

Q3 – Do instruments read the peak voltage at the top of the Sine Wave?

A3 – NO: All instruments from Voltmeters to your Electric Meter at home read the RMS (Root Mean Squared) voltage which is 0.707 of the peak.

A4 – How many phases are there?

Q4 – THREE: We actually generate three phase, 120 degrees apart making 360, which is transformed up and down by transformers in substations. Used to be Red, Yellow & Blue now Brown, Black & Grey. It is transmitted as three phase throughout the country at various voltages the largest being around 450,000 Volts. Industry will use the three phases in various items of equipment i.e. Electric Motors throughout the factory. However domestic premises will usually only use one of the phases plus a neutral which is created at the distribution transformer. (Brown and blue)

Q5 – We have talked about Voltage, where is the Current?

A5 – USUALLY LAGGING THE VOLTAGE: With an inductive system, i.e. transformers and motors. With so many coils of wire, the current is another smaller Sine Wave lagging the Voltage Sine Wave. If the system is capacitive then the current Sine Wave will lead the Voltage Sine Wave.

Usually we talk about inductive systems. The Cosine of the angle by which the current lags or leads the voltage is called the **Power Factor (Cos φ Phi)** which should be as near to **1 (Unity)** as possible. We can correct the power factor by inserting huge Inductors or Capacitors into the system. Industrial Power Factors may run at around 0.8.

Q6 – How can we change electric motor speed or lighting brightness?

A6 – CHOP THE SINE WAVE OR CHANGE FREQUENCY: We used to do this task by varying the voltage, but with motors this could result in reduced torque and dropping the voltage causes a problem with discharge lamps which includes fluorescents. These days we can do it with **Pulse Width Modulation.**

Q7 – Can we measure AC voltage?

A7 – NO: All instruments measure current and change to voltage on the scale for you.

DC Voltage:

DC Voltage is the most efficient when comparing it with an AC voltage. One of the main drawbacks is volt drop. DC cannot be transmitted over large distances like the national grid. They have just opened the longest ever cable run from Norway to Northumberland in the UK, it is around 450 miles long. It will power thousands of homes with green energy as Norway is 'hydro' (Water) power. If this cable run was DC I would hate to think what was left when it reached Northumberland. There are several types of DC let us have a look at some of them and this may give you the idea.

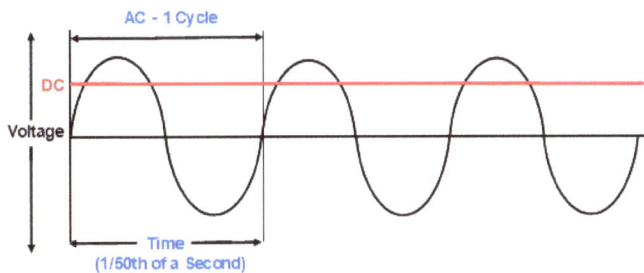

Changing from AC to DC is fairly easy. I have shown AC sine waves left and the DC in red. I have shown the DC as 0.707 of the sine wave which is the **'Root Mean Squared'** (RMS) value (explained in a previous section). So the straight line of DC is what we are after as our most efficient voltage so how do I obtain it?

Well, I could get a battery, the output of which would be DC straight line similar to the red line above and I could not have a better example. The voltage would depend upon my battery. So if I want to create DC another way, say, from AC, then that would be a little bit more involved so how?

If I take the AC sine wave (above) and put a diode into the supply. As the AC voltage goes first positive and then drops down to negative and then positive again, putting the diode in would stop the sine wave from going negative hence I would have a type of DC be it very 'noisy' as per the diagram on the right. This would be 'half wave'.

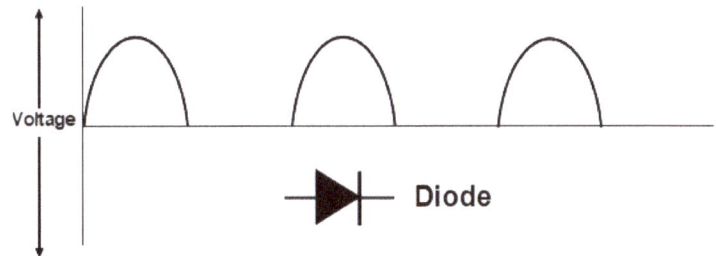

What do I mean by 'noisy' well there are gaps in the distance between the peaks so there would be a peak every $1/50^{th}$ of a second and then a gap before the second one etc.

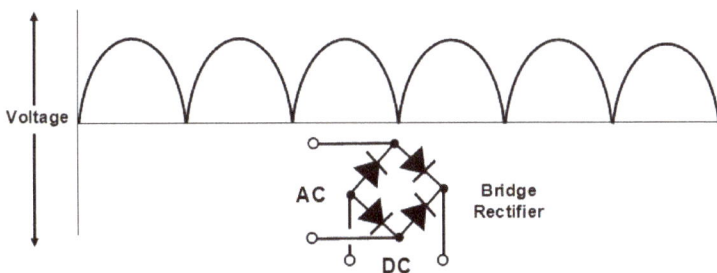

By using a bridge rectifier (shown left) I can invert the negative part of the sine wave up to the positive to fill in the gaps which would be less 'noisy'. As you can see we are now getting closer to a straight line which would, as in the case of the battery, be our ideal DC output. This would be 'full wave'.

So can we improve on the above to make our DC even smoother?

The answer is yes as I have shown below:

The top red line is our battery. By adding a capacitor to the circuit I can fill in the valleys and this may be as good as I can achieve. This DC would be quite smooth and could be used for my DC equipment, including instruments. I hope that this page has shown how we can use rectification and capacitors to obtain our DC.

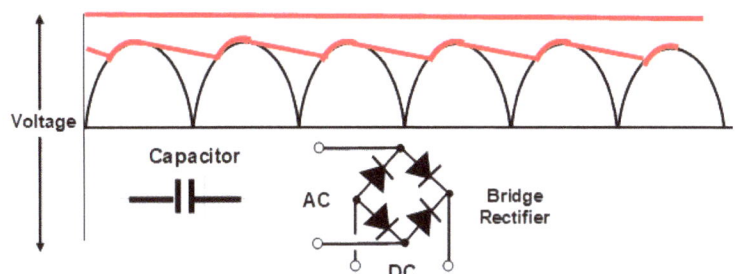

Sum up Questions on DC Voltage:

Q1 – Is DC Voltage more efficient than AC Voltage?

A1 – **YES:** In AC the voltage is less efficient as it is constantly passing through zero.

Q2 – Are there phases with DC?

A2 – **NO:** All one system similar to single phase in AC.

Q3 – Can I change DC to AC?

A3 – **YES:** With a piece of equipment called an Inverter.

Q4 – Can I change AC to DC?

A4 – **YES:** We do this easily and quite regularly with our battery chargers. Generators that produce AC can be fitted with a commutator to change output to DC.

Q5 – Will transformers work on DC?

A5 – **NO:** Transformers need the induced voltage of the primary to cutting the coils in the secondary at 50Hz to enable them to work. DC voltage can be dropped using equipment such as a rheostat.

Q6 – Will AC electric lamps work on DC?

A6 – **CERTAIN ONES:** If they are incandescent lamps and car or torch bulbs, i.e. they have a metal filament. If they are discharge lamps then, **NO**. Remember that here we have a ballast controlling the current in an arc tube.

If we look at AC equipment plugged into DC that is totally different. Because there is no frequency there is no inductive reactance, so it may see the AC equipment with its coils as a short circuit and burn out. It is possible to obtain universal equipment for AC or DC (Universal Motor).

Q7 – Why not use DC instead of AC?

A7 – **AC IS EASIER TO TRANSMIT & TRANSFORM:** AC can be transmitted over very long distances and can be transformed up and down quite easily. Transformers will of course not work on DC, so controlling it would be extremely difficult and any distance would cause volt drop. DC motors are used on equipment such as cranes, as we have much more direct control. DC is also used on car electrical systems, children's electric toys, torches etc. courtesy of the battery. DC is used on most electronic circuits.

Q8 – Is there such a thing as a Power Factor with DC?

A8 – **NO:** There is no frequency involved with DC, so the current will always be in phase with the voltage. I have said 'NO' but we could say that the power factor is always in unity with DC as current is always in phase with the voltage.

Q9 – Is Static Electricity DC?

A9 – **YES:** Including lightning, which of course is very high voltage Static Electricity.

Q10 – Can I measure voltage?

A10 – **NO:** All instruments measure current and change it to voltage on the scale for you.

Incandescent Lamps:

Incandescent Lamps are lamps where the light is obtained from a glowing element, whereas in discharge lamps, the light comes from an arc in a discharge tube.

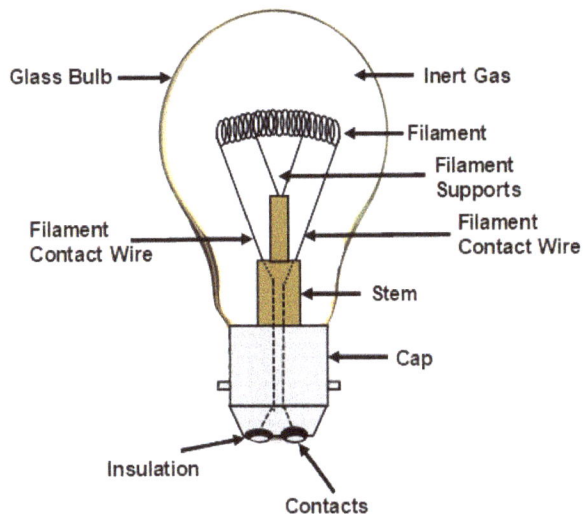

The old lamp (left) is called an '**Incandescent Lamp**' and is the one that has been used in homes through-out the UK for years. The inert gas inside of these lamps is mainly **Argon** with a small amount of **Nitrogen**. The filament is **Tungsten** which evaporates when the lamp is working, although argon will slow this process, and if positioned cap down, as shown in the diagram left, evaporated Tungsten collects as a black deposit on the inside top of the glass bulb. Eventually you would switch on and the lamp would blow, because when the element is cold that is the point of least resistance and maximum current and there would not be enough thickness of Tungsten to support the current.

These lamps very rarely go whilst working. The black insulation at the base which keeps the electrodes apart is a substance called **Vitrite**. The Tungsten filament has to be installed in one long continuous length as it cannot be joined because the filament heat is a few thousand degrees Centigrade when lit, and as no joint will stand that temperature the filament has to be coiled and coiled again and again to get in the amount of Tungsten to stand the current. The filament is held in place by insulated filament supports coming out of a glass stem and fed with live and neutral through the insulated stem from the cap and contacts at the bottom. Very old incandescent lamps were called **Nernst Lamps** which used ceramic rods which heated to huge temperatures instead of Tungsten, and these lamps did not require to be filled with inert gases.

The lamps could be obtained with different caps of: Bayonet cap (BC), Small Edison Screw (SES), Standard Edison Screw (ES) and Large Edison Screw (LES). With Edison Screw lamps, for safety, it is important that the cap thread is the neutral and the contact in the middle is the live. If the screw thread is the live then anyone fitting the lamp in, and who was in contact with the metal thread would get an electric shock if the power was switched on.

Manufacturers will state whether the lamp has to be '**Vertical Cap Down**' as the diagram at the top, '**Vertical Cap Up**', '**Horizontal**' or '**Universal**'. If the incorrect lamp is used in the wrong position then this could dramatically cut down the life of the lamp. Incandescent lamps like to be vertical cap up or vertical cap down. If they are horizontal then they are likely to fault quicker. Apparently it is to do with the filament supports, as these supports are thin wires with a coiled loop at the end holding the filament in place. If the lamp is horizontal the element is likely to rest on the part where the loop twists and hence rests on a double wire instead of a single and this causes a '**cold spot**' on the filament which is weaker than the rest. In chandeliers of the past, where the lamps stuck out sideways you would usually see one or two blown.

So why take this lamp completely off the market after being the most common lamp in history and change to others such as LED? Well this lamp is **VERY** energy inefficient, most of the lamp's energy, **around 90%** would you believe, is given out in **HEAT** rather than **LIGHT, which is around 10%**. We can now obtain lamps that will give out more light and much less heat and hence use much less energy. When you look at how many millions of this type of lamp were used in all of the millions of households in the UK, then using these old incandescent lamps could amount to quite a lot of wasted energy in heat rather than light which is why we are changing to LEDs. There are still uses for incandescent lamps where heat is required as well as light i.e. Grow lights, reptile tanks and medical Infra-Red lamps. Incandescent lamp life span is around 1000 hours.

Sum up Questions on Incandescent Lamps:

Q1 – Can I use a standard dimmer on an Incandescent Lamp?

A1 – YES: With a wire filament it will be okay to chop the voltage and the lamp will become dimmer or brighter as a result. Leading or Trailing Edge Dimmer should be okay here.

Q2 – Will Incandescent Lamps work with DC?

A2 – YES: They should work with AC or DC - perhaps not the same Lumens output.

Q3 – Can I just change my Incandescent Lamp for an LED Lamp?

A3 – YES: In fact the LED Lamp will be a much lower Wattage than the Incandescent Lamp you are taking out for the same Lumens output. (Better Efficacy)

Q4 – Are Incandescent Lamps efficient?

A4 – NO: An Incandescent Lamp is a lamp with a filament as against an arc. 90–95% of the input energy is given out as heat. Only 5-10% of the energy is given off in light which is why they are being phased out. Just think how many households there are in the UK and how many lamps/bulbs per household? Look at the amount of energy that is being wasted!

Q5 – What is the black on the inside of the glass on an Incandescent Lamp?

A5 – TUNGSTEN EVAPORATION: Incandescent Lamps have wire filaments as against discharge lamps, and when the lamp is lit the Tungsten Filament at 3000ºC is constantly evaporating and collects as a black deposit, on the inside of the glass bulb noticeable especially if the lamp is installed vertical cap down.

Q6 – Why do Incandescent Lamps sometimes blow when you switch on the light?

A6 – POINT OF MAXIMUM CURRENT & LEAST RESISTANCE: When you switch on the light with a cold filament, it is the point of least resistance and maximum current. As answer to question 5, if the Tungsten evaporates too much there is not enough left to sustain the current and the lamp will blow. They very rarely blow when the lamp is working normally.

Q7 – Will Incandescent Lamps blow quicker if they are installed horizontal as in some chandeliers, rather than vertical cap up or down?

A7 – SOMETIMES: It all depends upon the filament support wires. These are what hold the filament in position. The part which is in contact with the filament is just a twisted loop of wire. If the filament rests on the loop where it is double twisted this will cause a 'cold spot' on the filament and it could blow prematurely.

Chandeliers that have their lamps vertical cap up or down will not have this problem as the filament will always rests on one turn of the support wire so the lamp will go to the end of its natural life span.

Q8 – Can I throw Incandescent Lamps in the dustbin?

A8 – DEPENDS UPON RECYCLE LAWS: A very tricky question. What material are in the lamp? **Glass** – Outer bulb & inner supports, **Vitrite** – Black glassy solid insulator at base of cap, **Tungsten** – Filament, **Copper** – contact wiring, **Lead** – Very small amount for contacts at the base, **Argon & Nitrogen** – Gases inside of the bulb. **The good news is that this lamp has no Mercury.**

Halogen Lamps:

Some seem to think that Halogen Lamps and Metal Halide Lamps are the same thing, but let me now say that they are quite different. This Halogen lamp is an Incandescent Lamp not a Discharge Lamp meaning that the light comes from a glowing metal filament not a gas in a discharge tube.

Halogen:	Periodic Table:	Atomic Number:	At Room Temperature:	Colour:	Boiling Point:	Main Uses:	
Astatine	At	89	Solid	Black	337C	Radioactive - Medical	
Bromine	Br	35	Liquid	Dark Red	59C	Used in Lamps	
Chlorine	Cl	17	Gases	Pale Green	-34.6C	PVC	
Fluorine	F	9	Gases	Pale Yellow	-188.1C	Drinking Water	
Iodine	I	53	Solid	Black	184C	Used in Lamps	

The table above shows that there are five halogens, but only two are used in lighting. As you can see the boiling points are quite high, which is why the capsule has to get so hot. It is obvious that a lot of energy is lost in the way of heat and not much is used to produce light which is another reason why **Light Emitting Diode Lamps, LEDs, are taking over!**

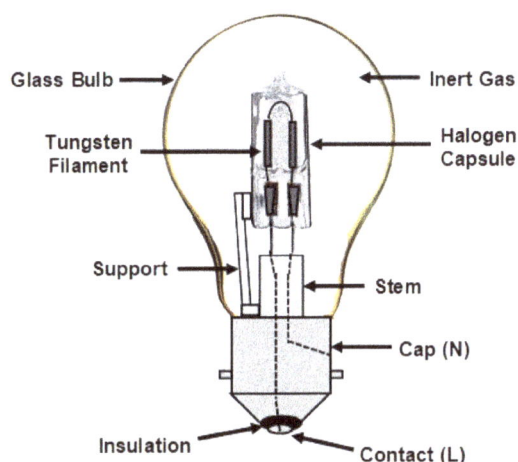

The lamp, shown here in the diagram (left), is a **Quartz Halogen Capsule** suspended within a glass bulb. The Quartz capsule has a Tungsten filament hence it may be called a **Quartz Tungsten Halogen Lamp**. As well as a similar design to Incandescent Lamps, these can be used in flush ceiling lights, projectors, car headlights etc. This lamp manufacture was due to be phased out by the end of 2021. Voltages range from quite low, 12 Volts & 24 Volts to 240 Volts. Again, these will be replaced by LED lamps. As below, if in just quartz capsule form without the outer bulb you must not handle as the oils in your skin will affect the Quartz. Use the packet to handle as a buffer between fingers & lamp.

Because of the very high pressure and temperature, if the capsule was made out of pure glass the heat would destroy the glass. With the capsule being made of quartz (fused silica) and not pure glass it is very sensitive about being handled. Your fingers can deposit oils out of your skin onto the quartz and cause a hotspot on the surface and so premature capsule rupture when it heats up. Capsules should be thoroughly cleaned if handled. Another problem, as mentioned above, is heat. Being a Tungsten filament producing the light, much of the energy is wasted in heat. The capsule would have to reach around 200°C to boil the Iodine Halogen.

When the power is switched on the Tungsten filament will glow inside of the Halogen capsule and heat up the Halogens, i.e. Iodine or Bromine, to their boiling temperature and the light is emitted almost instantaneously. In a normal Incandescent Lamp the Tungsten atoms evaporate when the lamp is lit and collects as a black deposit on the inside of the glass bulb, but in this case the Tungsten atoms, when the capsule cools down, because of the chemical reaction, collect back onto the element and as a consequence this lamp will last much longer than say an normal incandescent lamp. This is known as the **Tungsten Halogen Cycle**.

Sum up Questions on Halogen Lamps:

Q1 – Can I use a standard dimmer on an Incandescent (Halogen) Lamp?

A1 – YES: It should be okay to chop the voltage with a wire filament and the lamp will become dimmer or brighter as a result.

Q2 – Will Incandescent (Halogen) Lamps work with DC?

A2 – YES: They should work ok with AC or DC perhaps not the same lumens output.

Q3 – Can I just change my Incandescent (Halogen) Lamp for an LED Lamp?

A3 – YES: In fact the LED Lamp will be a much lower Wattage than the Incandescent Lamp you are taking out for the same Lumens output. (Better Efficacy)

Q4 – Are Incandescent (Halogen) Lamps inefficient?

A4 – YES, VERY: In an Incandescent (Halogen) Lamp 90% of the input energy is given out as heat. Only 10% of the energy is given off in light, which is why they are being phased out.

Q6 – Does the Tungsten filament evaporate here on the Incandescent (Halogen) Lamp like a normal standard Incandescent Lamp?

A6 – TUNGSTEN WILL EVAPORATE: When the lamp is lit the Tungsten filament is constantly evaporating, but in this case of the Halogen Lamp because of the chemical reaction, it collects back onto the filament not the Quartz glass.

Q7 – Can I throw Incandescent (Halogen) Lamps in the dustbin?

A7 – DEPENDS UPON RECYCLE LAWS: A very tricky question. What materials are in the lamp? **Quartz Glass** – Capsule, **Glass** – Outer bulb & inner supports, **Vitrite** – Black glassy solid insulator at base of cap, **Tungsten** – Filament, **Copper** – contact wiring, **Lead** – Very small amount for contacts at the base, **Bromine or Iodine** – Gases inside of the capsule. **The good news is that this lamp has no Mercury.**

Q8 – Is the Halogen Capsule made out of pure glass?

A8 – NO: It is made out of Quartz Fused Silica. Because of the very high pressure and temperature, if the capsule was made out of pure glass the heat would destroy the glass.

Q9 – How hot does the filament get on an Incandescent (Halogen) Lamp?

A9 – VERY HOT! Around 3000ºC.

Q10 – How hot does the Quartz Capsule get on an Incandescent (Halogen) Lamp?

A10 – VERY HOT! The capsule would have to reach around 200ºC to boil a Halogen and cause the chemical reaction to stop Tungsten collecting on the inside of the Quartz Capsule.

Q11 – Can I touch the Quartz Capsule with by fingers?

A11 – NO: Your fingers can deposit oils out of your skin onto the Quartz Fused Silica and cause a hotspot on the surface and so premature capsule rupture when it heats up.

Capsules should be thoroughly cleaned if handled! Use the packet as a shield when replacing a Capsule!

Mercury Vapour/Mercury Blended Lamp:

This is a **High Intensity Discharge Lamp (HID)** (discharge lamps have an arc tube and rely on gas discharge for light instead of a glowing element) and consists of a glass bulb filled with Nitrogen. Inside of the bulb is a support frame (red dotted in the diagram below) holding a glass arc tube filled with Argon & Mercury gas, which may reach very high temperatures when working. This arc tube is where the light is emitted and is achieved by two tri-metallic electrodes, dipped in Thorium, Calcium and Barium Carbonates. I have renewed hundreds as a young electrical technician. We called them Mercury Blended lamps. They are due to be phased out when present stocks are exhausted.

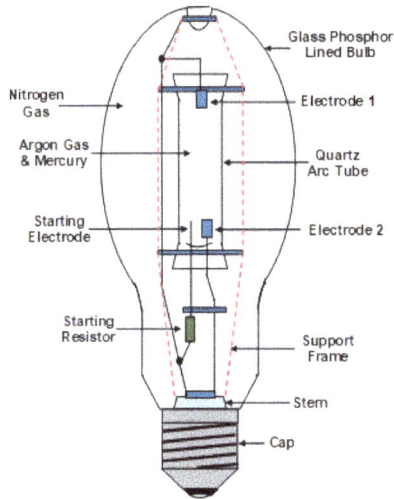

Diagram labels:
- Glass Phosphor Lined Bulb
- Nitrogen Gas
- Electrode 1
- Argon Gas & Mercury
- Quartz Arc Tube
- Starting Electrode
- Electrode 2
- Starting Resistor
- Support Frame
- Stem
- Cap

When the power is applied there is a higher voltage causing a local arc between the starting electrode/starting resistor and the bottom main electrode 2. As the Argon gas is ionised by the arc, the current is being limited at this point by the starting resistance. This starting arc heats up and vaporises the small amount of liquid Mercury causing the main arc to flash from the bottom electrode No.2 to the top electrode No.1. From switching on to full light will take around 5 minutes. If the arc was to fail to strike then the lamp goes out and it all starts all over again. This is known as the lamp being '**in cycle**' and the lamp must be changed. Like a fluorescent lamp once the arc forms there is more or less a short circuit so a ballast is required to limit the current otherwise the arc tube would burst. On industrial lamps the ballast is a separate unit in the base. Some lamps may have a built in ballast. Ensure the lamp size matches the ballast.

As the gas in the arc tube heats up, the pressure and temperature within the tube increases and if the arc tube was made of glass instead of Quartz the high temperature & pressure would destroy the glass. The outer glass, a Nitrogen filled bulb, can either be coated with white phosphor or clear and provides thermal insulation as well as protection from the intense ultra-violet which the arc will produce. The light is initially bluish white in colour and in clear lamps it remains this colour, but if internally coated with phosphor it may correct this to a warmer light. A detrimental feature is being very bright around the lamp itself, but it can appear to be darker on the ground directly under the lamp. The lamp also contains shock absorber springs in case of rough handling.

The efficacy, how many watts in compared to how much light out, is very good: something in the region of 30-60 Lumens/Watt which is much better than incandescent lamps. Lifespan is around 25,000-100,000 hours. A drawback is that if the supply is interrupted then the lamp obviously goes out and has a restrike time of several minutes to get to full brightness again. As above, some lamps are what is called, self-ballasted, and have an extra element, but others may require a choke/ballast to limit the current similar to a fluorescent once the arc is struck. Never used in emergency lighting because of the restrike time.

Unfortunately, as the name suggests, the lamp contains Mercury and is therefore not environmentally friendly. Faulty lamps have to be disposed of in lamp disposal units and not just thrown in dustbins or skips. Breakage of the Nitrogen filled outer bulb whilst the lamp is lit can expose the intense ultra-violet of the inner arc tube and many lamps have a power 'cut out' built in, which activates after so many seconds should this happen. These lamps started as low pressure Mercury lamps and later were designed and manufactured as high pressure lamps. Also ensure that the lamp is installed per design i.e. vertical cap up, down or horizontal, otherwise its lifespan will be cut dramatically. Many lamps are universal. See below for installation details:

These lamps have letters e.g. **MBFTU - M** = Mercury Discharge Lamp, **B** = Quartz Envelope loaded below 100W/cm of arc length, **F** = Fluorescent outer bulb, **T** = Incandescent Tungsten filament **U** = Universal cap. **Other letters may include C** = Glass Envelope loaded below 10W/cm of arc length, **D** = Quartz Envelope with forced liquid cooling, **E** = Quartz Envelope loaded above 100W/cm of arc length, **V** = Vertical Cap up, **D** = Vertical Cap down, **H** = Horizontal cap, **W** = Wood's Glass.

Sum up Questions on Mercury Discharge Lamps:

Q1 – Can I use a standard dimmer on Discharge Mercury Vapour/Blended Lamps?

A1 – IT IS POSSIBLE BUT NOT COMMON: If we look at this with the electrical knowledge that we have and the information previously shared, discharge lighting has no metal filaments like an Incandescent Lamp or Metal Halide Lamp, so varying the voltage or frequency here could be catastrophic or at least producing inefficient light.

We have a ballast and its job it is to keep down the current in the arc tube, if we drop the voltage that current in theory, could in fact go up and destroy the arc tube. We may investigate altering the waveform using Pulse Width Modulation. Manufacturers will give advice.

Q2 – Will Discharge Mercury Vapour Lamps work with DC?

A2 – NO: Remember that in many cases we are not just considering the lamp, but also the ballast.

Q3 – Can I just change my Discharge Mercury Vapour Lamp for an LED Lamp?

A3 – YES: In fact the LED Lamp will be a much lower Wattage than the Incandescent Lamp you are taking out for the same Lumens output. (Better Efficacy) Just remember that with an LED Lamp the Discharge & Fluorescent Lamp ballast will not be required. **These lamps are due to be phased out when present stocks exhausted.**

Q4 – Are Mercury Discharge Lamps efficient?

A4 – YES: Excellent efficacy (power in to output ratio) and life span. Unfortunately they contain Elements that are undesirable, such as Mercury, Sodium etc.

Q5 – What is the white powder on the inside of the Mercury Discharge Lamp?

A5 – PHOSPHOR: This is a toxic white powder coating which gives off a desirable light when 'Photons' bounce off it.

Q6 – Will Mercury Discharge Lamps blow quicker if they are installed horizontal when manufacturers state that they should be vertical and visa-versa?

A6 – MOST DEFINITELY: The manufacturers will provide instructions, sometimes with letters on the glass of the lamp, as to whether it should be mounted Vertical Cap Up, Vertical Cap Down, Horizontal or Universal. Mounted in the wrong plane will cut the lifespan of the lamp enormously.

Q7 – Can I throw Mercury Discharge Lamps in the dustbin?

A7 – DEFINITELY NOT! As the name suggests they contain Mercury and require specific disposal.

Q8 – Can I change Mercury Discharge Lamps for another i.e. Sodium?

A8 – NO: You not only have to consider the Discharge Lamp itself, but also its ballast which would not in many cases be compatible. Consult Manufacturers.

Q9 – Why is a Mercury vapour lamp called a negative resistance lamp?

A9 – RESISTANCE KEEPS DECREASING AS THE CURRENT INCREASES: In theory the current could increase so much that the lamp would destroy itself, hence there is a ballast to limit the current.

Metal Halide Lamps:

This is a **High Intensity Discharge Lamp (HID)** meaning it has an arc tube and relies on gas discharge for light instead of a glowing element. Halides are a compound of a Halogen atom and another element to form another substance e.g. a salt.

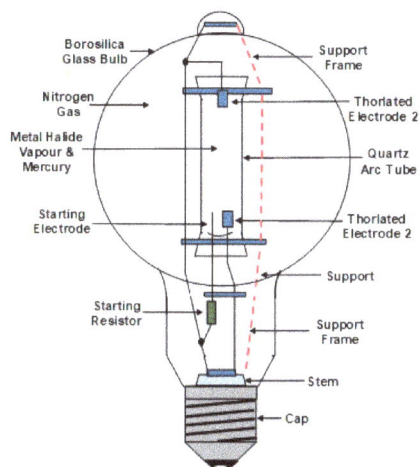

Borosilica Glass Bulb — Support Frame

Nitrogen Gas — Thorlated Electrode 2

Metal Halide Vapour & Mercury — Quartz Arc Tube

Starting Electrode — Thorlated Electrode 2

— Support

Starting Resistor — Support Frame

— Stem

— Cap

The Metal Halide Lamp (MH) is very similar to the Mercury Blended Lamp and consists of a glass bulb filled with Nitrogen. Inside of the bulb is a support frame (red dotted in the above diagram) holding a Quartz arc tube this time filled with Metal Halide Vapours as well as Argon & Mercury gas, which may reach very high temperatures when working.

On later versions of this lamp the arc tube was made from ceramics rather than Quartz, as they are more resistant to the effects of metal halides. These lamps were called **'Ceramic Metal Halide'** lamps. The arc tube is where the light is emitted and is achieved by two main electrodes and a starter electrode.

When the power is applied, there is a higher voltage causing local arc between the starting electrode and the bottom main electrode 2 as the gas is ionised, the current being limited at this point by the starting resistance. This starting arc heats up and vaporises the small amount of liquid Mercury causing the main arc to flash from the bottom electrode 2 to the top electrode 1. Once the main arc is formed the starting arc quenches. From switching on to full light will take around 5 minutes.

Halides are chemical compounds of Halogens, namely: Astatine, Bromine, Chlorine, Fluorine and Iodine. As the gas in the Quartz arc tube heats up and the pressure and temperature within the tube increases, if the arc tube was made of glass the high temperature & pressure would destroy the glass.

The outer glass is made of **Borosilicate (Pyrex)**, Nitrogen filled bulbs can either be coated with phosphor or clear and provides thermal insulation as well as protection from the intense ultra-violet which the arc will produce. The lamp contains shock absorber springs in case of rough handling.

The efficacy (how many watts in compared to how much light out) is very good, something in the region of 100 Lumens/Watt much better than incandescent lamps. Lifespan is also quite long, around 15,000 hours. A drawback is that if the supply is interrupted then the lamp obviously goes out and has a restrike time of several minutes to get to full brightness again. Some lamps are what is called self-ballasted and have an extra element, but others may require a choke/ballast to limit the current similar to a fluorescent once the arc is struck. **Never used in emergency lighting because of the restrike time.**

Unfortunately, the lamp contains Mercury and is therefore not environmentally friendly, **Hydrargyrum Hg (former redundant Greek name for mercury)**, and faulty lamps have to be disposed of in lamp disposal units and not just thrown in dustbins or skips. Breakage of the Nitrogen filled outer bulb whilst the lamp is lit could expose the intense ultra-violet of the inner arc tube and many lamps may have a power 'cut out' built in which activates after so many seconds should this happen. Also you need to ensure that the lamp is installed per design i.e. vertical cap up, down or horizontal otherwise its lifespan will be cut dramatically. Many lamps are universal.

Quartz Iodide Lamps are part of the Metal Halide family of lamps. These are used in decorative lamps that look like a huge translucent, usually orange crystal lit from the inside. (HMi, HQ & HQI Lamps).

Sum up Questions on Metal Halide Lamps:

Q1 – Are Halogen Lamps and Metal Halide lamps the same?

A1 – **DEFINITELY NOT:** A Halogen Lamp is an Incandescent Lamp meaning it produces light from a heated metal filament, a Metal Halide Lamp is a Discharge Lamp meaning the light is produced by an arc in a discharge tube.

Q2 – What are Metal Halides?

A2 – **COMPOUNDS OF HALOGENS:** Halides are a compound of a Halogen Atom and another Element to form another substance e.g. a salt.

Q3 – Can I use a standard dimmer on Discharge Metal Halide Lamps?

A3 – **IT IS POSSIBLE BUT NOT COMMON:** If we look at this with the electrical knowledge that we have and the information given, discharge lighting has no metal filaments like an Incandescent Lamp or Halogen Lamp. So varying the voltage or frequency here could be catastrophic or at least producing inefficient light.

We have a ballast and its job it is to keep down the current in the arc tube, if we drop the voltage that current in theory could in fact go up and destroy the arc tube. Again we may investigate altering the waveform using Pulse Width Modulation. Manufacturers will give advice.

Q4 – Will Discharge Metal Halide Lamps work on DC?

A4 – **NO:** Remember that in many cases we are not just considering the lamp, but also the ballast.

Q5 – Can I just change my Discharge Metal Halide Lamp for an LED Lamp?

A5 – **YES:** In fact for the same Lumens output, the LED Lamp will be a much lower wattage than the Incandescent Lamp you are taking out (better efficacy). Just remember that with an LED Lamp the Discharge & Fluorescent Lamp ballast will not be required.

Q6 – Can I change one type of Discharge Metal Halide Lamp for another i.e. Mercury for Sodium?

A6 – **NO:** You not only have to consider the Discharge Lamp itself, but also its ballast which would not, in many cases, be compatible.

Q7 – Are Discharge Metal Halide Lamps efficient?

A7 – **YES:** Excellent efficacy (power in to output ratio) and life span. Unfortunately they contain Elements that are undesirable such as Mercury etc.

Q8 – Will Discharge Metal Halide Lamps blow quicker if they are installed horizontal when manufacturers state that they should be vertical and visa-versa?

A8 – **MOST DEFINITELY:** The manufacturers will give instructions, sometimes with letters on the glass of the lamp as to whether it should be mounted Vertical Cap Up, Vertical Cap Down, Horizontal or Universal. Mounted in the wrong plane will cut the lifespan of the lamp enormously.

Q9 – Can I throw Incandescent Metal Halide Lamps in the dustbin?

A9 – **DEFINITELY NOT!** They contain Mercury amongst other things.

Low Pressure Sodium Lamps:

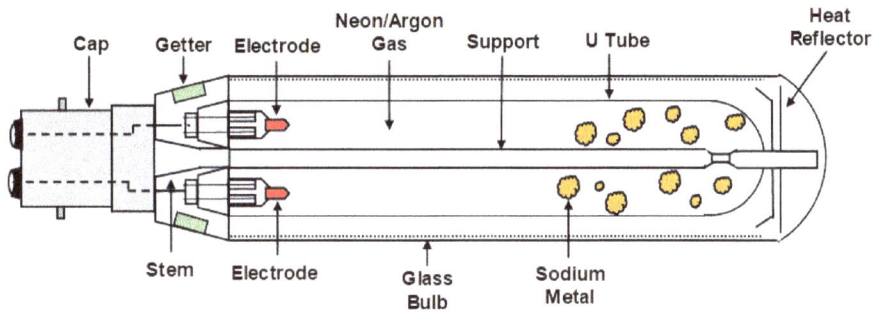

Not suitable for chemical factories as if the lamp is dropped or broken and Sodium Metal hits damp air it will explode! Low Pressure Sodium Lamps (LPS) are tubular in design, as in the above diagram, and may be called a **'SOX'** Lamp. These lamps when lit give off a yellow light as is usual with Sodium lamps. As you can see there is a borosilicate glass 'U' tube full of Neon and Argon gas and solid lumps of Sodium metal with an electrodes at each end. The whole discharge assemble is encased inside a glass outer bulb, which is a vacuum for thermal insulation, coated on the inside with **'Indium Oxide'** to protect us against the intense ultra-violet. This is a discharge lamp and about a third of the input energy is converted to light, much more than an incandescent lamp.

There are various support frames which keep the tube suspended. There is a material called a **'getter'** suspended in the vacuum. This keeps the vacuum stable by absorbing unwanted gases. This can be a device like the one shown or a coating within the glass. The 'U' tube could be designed to be stretched out by bending into a 'U' to make the lamp longer, it would make it more practicable. The lamp is called Low Pressure Sodium because the tube pressure is lower than atmospheric.

When the lamp is switched on, a voltage appears between the Tungsten electrodes and the Neon. Having a low thermal conductivity, Neon is the first to ignite and the lamp will take on the appearance of a pink light. The Argon further assists the arc and at this point the Sodium vaporises and the lamp will take on the familiar yellow appearance. The ballast will limit the current once the arc is struck as in a fluorescent light. Neon will ignite first causing a pink glow.

Whenever light is produced from a gas such as this one, they are called **'Discharge'** Lamps and usually have a very good efficacy, i.e. the number of watts of electrical energy compared with light output, of around 150-200 lumens/watt. The life span of low pressure sodium lamps are usually around 20,000 hours. This lamp will require a choke and ignitor unit. The warm-up time from switch on to full brightness can be around 15 minutes which could be a drawback in power failures. Uses: Street lighting, Car Parks & Public Areas.

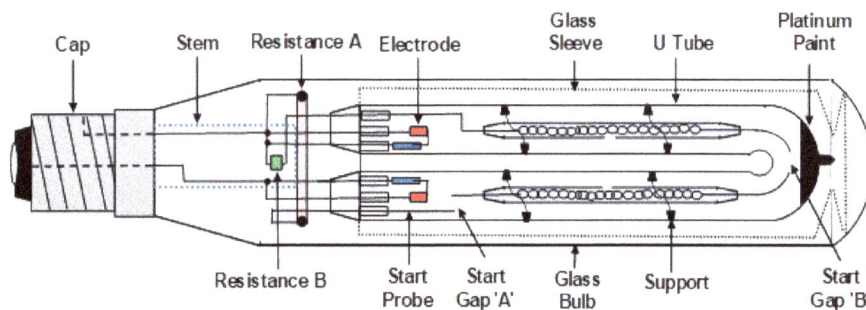

You must find out if your Sodium Lamp has an external or internal ignitor. This is done by looking at the glass for a triangle. A triangle with an **'I'** is internal ignitor and a triangle with an **'E'** is external ignitor. These are usually tubular lamps. Above is a diagram of a Sodium lamp with an internal ignitor. Mixing them up during installation can, in certain circumstances, be catastrophic and costly for the lamp. When these lamps are switched off there may still be a glow for several seconds as the Sodium is still at operating temperature this is called **'Ghosting'**.

Sum up Questions on Low Pressure Sodium Lamps:

Q1 – Can I carry a Low Pressure Sodium Lamp through a hazardous area?

A1 – DEFINITELY NOT! These are called 'SOX' Lamps, like you might see in yellow street lamps, and have a 'U' tube containing the Sodium Metal. If you were to drop the lamp and it was a damp day, the Sodium Metal would explode on contact with damp air.

Q2 – Can Low Pressure Sodium Lamps be obtained with internal and external ignitors?

A2 – YES: And you must not mix them up, or else their life will be shortened or they will simply not work at all. There should be an **'E'** in a triangle on the glass if the lamp requires an external ignitor and an **'I'** in a triangle if the lamp has an internal ignitor. They must be installed in the plane that the manufacturers have designed them for i.e. horizontal, vertical cap up or down or universal otherwise their life span will be cut short enormously.

Q3 – Can I use a standard dimmer on Discharge Low Pressure Sodium Lamps?

A3 – My answer here would be NO - for the following reasons. If we look at this with the electrical knowledge that we have and the information previously presented, discharge lighting has no metal filaments like an Incandescent Lamp or Halogen Lamp so varying the voltage here could be catastrophic or at least producing inefficient light. We have a ballast and its job it is to keep down the current in the arc tube, if we drop the voltage that current in theory could in fact go up and destroy the arc tube. Again we may investigate altering the waveform using Pulse Width Modulation.

Q4 – Will Low Pressure Sodium Lamps work on DC?

A4 – NO: Remember, in many cases we are not just considering the lamp, but also the ballast.

Q5 – Can I just change my Low Pressure Sodium 'SOX' Lamp for an LED Lamp?

A5 – YES: In fact the LED Lamp will be a much lower wattage than the Incandescent Lamp for the same Lumens output (Better Efficacy). Remember, with an LED Lamp the Discharge Sodium Lamp ballast will not be required. If you look at modern street lighting which is LED Lamps, the lamp units are much thinner.

Q6 – Can I change one type of Discharge Low Pressure Sodium Lamp for another e.g. Mercury?

A6 – NO: You not only have to consider the Discharge Lamp itself, but also its ballast which would not, in many cases, be compatible.

Q7 – Are Discharge Low Pressure Sodium Lamps efficient?

A7 – YES VERY! Excellent efficacy and life span is comparable to LED.

Q8 – Can I throw Low Pressure Sodium Lamps in the dustbin?

A8 – DEFINITELY NOT! As above the 'U' Tube contains Sodium Metal it could easily cause a fire in a dustbin or skip if it was broken and came into contact with any damp air or water. They should be disposed of in a lamp disposal unit.

Q9 – What is a 'Getter'

A9 – A SUBSTANCE THAT REMOVES IMPURITIES.

High Pressure Sodium Lamps:

This is a **High Intensity Discharge Lamp (HID)** meaning it has an arc tube and relies on gas discharge for light instead of a glowing element. The arc tube pressure being higher than atmospheric.

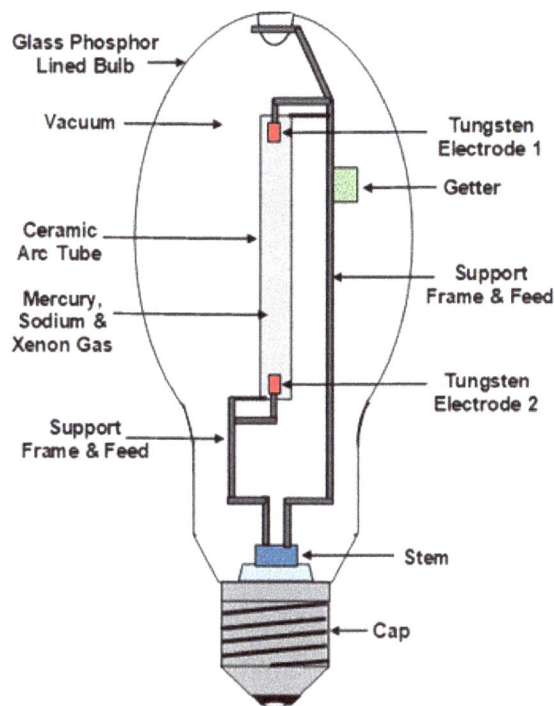

Glass Phosphor Lined Bulb →

Vacuum →

← **Tungsten Electrode 1**

← **Getter**

Ceramic Arc Tube →

Mercury, Sodium & Xenon Gas →

← **Support Frame & Feed**

← **Tungsten Electrode 2**

Support Frame & Feed →

← **Stem**

← **Cap**

High Pressure Sodium Lamps (HPS) come in a variety of shapes and sizes. I have shown an elliptical style lamp in the diagram left, but tubular ones are quite common. These lamps, when alight, give off a yellow/orange light as is usual with Sodium based & Mercury lamps. As you can see there is a translucent ceramic arc tube full of Xenon Gas and a Sodium & Mercury amalgam with a Tungsten electrode at each end. This translucent tube, made mainly out of **Aluminium Oxide (Alumina)**, has to withstand extreme high temperatures when the lamp is working, as well as the very corrosive effect of Sodium. Sodium would attack a tube made out of glass. The outer bulb can be clear or coated with phosphor.

You must find out if your Sodium lamp has an external or internal ignitor and this is done by looking at the glass for a triangle. A triangle with an 'I' is internal ignitor and a triangle with an 'E' is external ignitor These are usually tubular lamps.

Manufacturers catalogue their lamps differently. For instance, you can obtain SON-E and SON-T, the SON-E is elliptical and SON-T is tubular. We had many of these on our complex when we were replacing the Mercury Vapour Lamps. Any change of type of lamp in an Atex certified fitting requires manufacturer approval **in writing (i.e. Certificate of Conformity)**.

The tube is encased in a glass bulb with a vacuum medium. There are various support frames which keep the tube suspended. There is a material called a **'getter'** suspended in the vacuum and this keeps the vacuum stable by absorbing unwanted gases. This can be a device like the one in the diagram above (Getter) or a coating within the glass. One of the initial problems they had was to completely seal this tube against the high pressure, a very corrosive Sodium Gas was used. They now use a **Monolithic** seal.

Let us have a look at internal ignition. When the lamp is switched on a voltage appears between the Tungsten electrodes and the Xenon, having low thermal conductivity, it is the first to ignite and the lamp will take on the appearance of light blue. Next, the Mercury vaporises and further assists the arc and the tube takes on a whitish blue look and at this point the Sodium vaporises and the lamp will take on the familiar yellow/orange appearance. Yellow is the Sodium and the Mercury vapour mix gives it an orange look. A ballast is used to control the current once the arc is struck.

Whenever light is produced from a gas such as this one, they are called **'Discharge'** lamps and usually have a very good efficacy i.e. the number of watts of electrical energy compared with light output, of around 100-150 lumens/watt. The life span of high pressure sodium lamps is usually around 25,000+ hours. Some HP Sodium lamps can be dimmed if they have a special ballast.

This lamp will require a choke and ignitor unit to send high voltage pulses using the ballast coil to the arc tube. The warm up time from switch on to full brightness could be around 15 minutes which could be a drawback in power failures. Uses: Street lighting, Car Parks & Public Areas. Also ensure that the lamp is installed per design i.e. vertical cap up, down or horizontal otherwise its lifespan will be cut dramatically. Many lamps are universal.

Sum up Questions on High Pressure Sodium Lamps:

Q1 – Can I carry and use High Pressure Sodium Lamp a hazardous area?

A1 – YES: These have not got the amount of Sodium Metal and are used extensively.

Q2 – What is the difference here to an Incandescent Lamp?

A2 – ARC VERSUS FILAMENT: Here the light comes from an arc inside of an arc tube whereas in an Incandescent Light it comes from a glowing metal filament.

Q3 – Can Low Pressure Sodium Lamps be obtained with internal and external ignitors?

A3 – YES: And you must not mix them up or else their life will be shortened or they will simply not work at all. There should be an 'E' in a triangle on the glass if the lamp requires an external ignitor and an 'I' in a triangle if the lamp has an internal ignitor. Also they must be installed in the plane that the manufacturers have designed them for i.e. horizontal, vertical cap up or down or universal otherwise their life span will be cut short enormously!

Q4 – Can I use a standard dimmer on Discharge High Pressure Sodium Lamps?

A4 – SOME OF THEM: In these cases you have not just got the lamp to consider, but also the ballast. So some HP Sodium Lamps can be dimmed if they have a special ballast.

Q5 – Will High Pressure Sodium Lamps work on DC?

A5 – NO: Remember that in many cases we are not just considering the lamp, but also the ballast.

Q6 – Can I change my High Pressure Sodium Lamp for an LED Lamp?

A6 – YES: In fact the LED Lamp will be a much lower Wattage than the Incandescent Lamp for the same Lumens output (Better Efficacy). Just remember that with an LED Lamp the Discharge Sodium Lamp ballast will not be required. If you look at modern street lighting which uses LED Lamps, the lamp units are much thinner. With Atex lighting Manufacturers must be consulted.

Q7 – Can I change one type of Discharge High Pressure Sodium Lamp for another e.g. High Pressure Mercury Lamp?

A7 – NOT WITHOUT CHECKING DATA: You not only have to consider the Discharge Lamp itself, but also its ballast which would not in many cases be compatible.

Q8 – Are Discharge High Pressure Sodium Lamps efficient?

A8 – YES: Excellent efficacy (power in to output ratio) and life span.

Q9 – Can I throw High Pressure Sodium Lamps in the dustbin?

A9 – DEFINITELY NOT! The Lamp as the name suggests contains Sodium and as such could easily cause a fire in a dustbin or skip if it was broken and came into contact with any damp or water. One of the other problems here is that they also contain Mercury, which is not good for the environment. They should be disposed of in a lamp disposal unit.

Q10 – What is the white coating on the inside of the glass bulb?

A10 – PHOSPHOR: A white toxic powder coating which emits a 'friendly' light when hit by Photons. Similar to a fluorescent tube.

Older Single Fluorescent Lamps (with Starter):

Looking at the diagram below, the AC supply is switched on which puts power onto the right-hand lower pin of the tube, which flows through the electrode/heater (Red) and out of the top right pin to the 'starter'. In the starter there is a gap between the bi-metallic strip and the circuit, which when the power is on an arc forms, which ionises the gas in the 'Starter'.

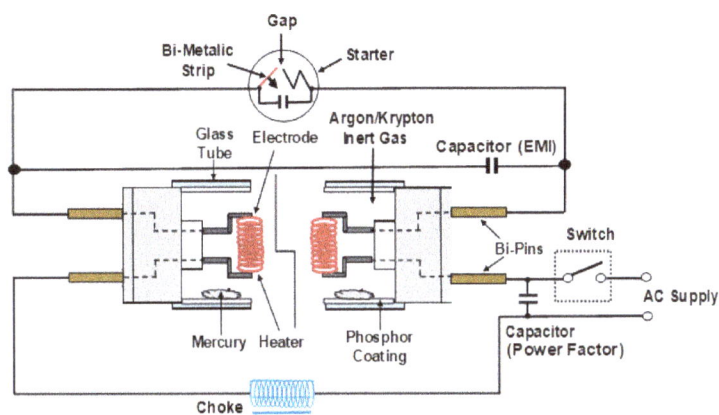

Because the 'Starter' has a bi-metallic strip it begins to operate due to ionised gas heating it up and it closes. This powers the heaters on the tube electrodes which heat up and warm the gas in the tube. It will glow at either end due to the ionised gas around the electrodes. Without the heaters preheating the gas, even with the high voltage, it may not be enough for an end-to-end connection When the bi-metallic strip touches the electrode in the starter this puts out the arc in the starter.

Power is then put onto the tube heaters/electrodes and a circuit is formed through the pins, through the starter and through the electrodes. At this point the ionised gas in the starter ceases because there is no arc and the bi-metallic strip cools down and opens a very high inductive circuit with the choke. At this point a high voltage pulse in the tube and the flashover takes place, and the small amount of Mercury in the tube will vaporise. If the tube does not flash over, the process in the starter will start all over again. The 'starter' has to fail open circuit so if the tube is flashing because of a faulty starter then by removing the starter then the tube will come on, but will not light again if switched off until the starter is replaced.

Being a discharge lamp, the arc in the main tube flashes from end to end and at this point there is virtually a short circuit; so something has to limit the current and that is the coils in the choke. Modern chokes are likely to be electronic and not the old fashioned coil type that I have drawn. Because of the Mercury vapour in the tube, ultraviolet light is produced which turns to visible light as the electrons emit photons which bounce off the white phosphor coating on their way from one end to the other. This process is called 'Thermionic Emission'. Without Phosphor there would be just Ultraviolet non visible light. The capacitor, which is in the fluorescent circuit across the main incoming supply, is there to correct the power factor. The coils of wire in the choke would cause an inductive load, so the current would lag the voltage. If you consider how many thousands of chokes there are, this would be a huge problem. The other larger capacitor is for Electromagnetic Interference (EMI)

If you look down, say, a corridor which is lit with fluorescent lighting, you will notice that the lights are dimmer when first switched on and get brighter as the Mercury gas in the tubes heats up. As the tubes age you will see different colour as the light intensity changes and the tubes will start to get black around the ends due. This is due to the evaporation of the Tungsten heater/filament collecting on the glass as well as other impurities. **Dimmers cannot be used on standard fluorescent lights as you cannot lower the voltage!**

Older Double Fluorescent Fittings:

The question arises, why do we use double tube fluorescent fittings in many occasions? In a workshop where there is rotating machinery such as lathes, milling machines and drills, it is really essential that the fluorecent lighting was double fluorescents. The tube flashes at 50Hz, and although this may not be noticed with the human eye, it is possible that the fluorescent flashing could act as a Stroboscope and make machines appear stationary. To counteract this an extra capacitor is fitted in one tube circuit so the tubes do not flash together. There is also more light from double fittings which cut down shadow effects (but not double the light as you may think!).

Sum up Questions on Fluorescent Lighting:

Q1 – What is the purpose of the starter?

A1 – HIGH VOLTAGE: If a coil circuit is opened and closed rapidly, a high voltage is produced to enable the arc to flashover from one end of the tube to the other. This is what happens here, the contacts in the starter open and close the choke/ballast circuit. The choke/ballast contains thousands of coils of wire.

Q2 – Can the above arc get out of control?

A2 – NO: Because the arc in the tube would, in theory, be a short circuit and another function of the choke/ballast is to enter the circuit and limit the current.

Q3 – Why does the tube glow at either end when the power is switched on?

A4 – GAS IONISATION: There is a heater at either end of the tube which heats and ionises the gas in the tube to assist the discharge. As the flashover occurs the Mercury in the tube will vaporise and the tube light will get brighter. (Hot Cathode)

Q5 – Why does the tube flash on and off?

A5 – THE STARTER: This starter has to fail open circuit when the tube is successfully lit, if it does not and keeps opening and closing, the tube will continue to flash. By removing the starter physically, the tube will come on as you have 'open circuited' it. The tube will not attempt to start again until the starter is replaced.

Q6 – Can I use a standard Leading Edge Dimmer on Fluorescent Lamps?

A6 – NO: Standard Fluorescent lighting does not like you to use a standard Leading or Trailing Edge Dimmer. I did say **NO**, but it is possible to dim fluorescent lighting, but it is a complex operation. You would end up with not enough energy for the tube arc, and the fitting would go into a mode where it is constantly trying to start. You may require a special dimming regulator along with special cable to the fitting. Ensure the fitting is dimmable in the first place as older fittings may not be. Contact manufacturers for advice.

Q7 – What is the white powder on the inside of the Fluorescent Tube?

A7 – PHOSPHOR: This is a toxic white powder coating which gives off a desirable light when 'Photons' bounce off it. Without a Phosphor coating there would just be an ultraviolet glow in the tube.

Q8 – Can I throw Fluorescent Tubes in the dustbin or skip?

A8 – DEFINITELY NOT: As well as toxic Phosphor they also contain Mercury.

Q9 – When the tube is at the end of its life why is it black on either end?

A9 – TUNGSTEN & IMPURITIES: At either end of the tube, combined with the heater units, are Tungsten Emitters (Electrodes) that are the point-to-point connections for the arc electrons called Thermionic Emission. The black is evaporated Tungsten, as well as other impurities. Once this black appears on the tube ends and the tube is coming to the end of its life, the emitted a large percentage of the light will be cut. Eventually, if left, the tube will just glow at either end and no arc will appear.

Q10 – What is the purpose of the small capacitor in a fluorescent fitting?

A10 – POWER FACTOR CORRECTION.

Compact Fluorescent Lamps (CFL):

This is a **'Discharge Lamp'**. Discharge Lamps have an arc tube and rely on gas discharge for light instead of a glowing element. It is based on a standard fluorescent tube but just smaller!

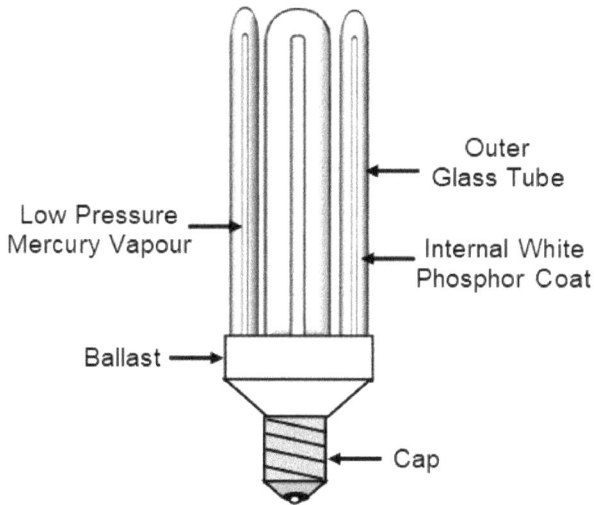

Outer Glass Tube

Low Pressure Mercury Vapour

Internal White Phosphor Coat

Ballast

Cap

The Compact Fluorescent Lamp (CFL) is used as a long life low energy consumption light in many homes and industries. They became popular in the 90s, but again one of the problems is that it contains Mercury, about 3-4 grams to be exact, which is not environmental friendly.

In industry these lamps should be disposed of in special lamp disposal units because of the Phosphor and Mercury but where do you dispose of them at home? Not in the dustbin I hope?

We replaced some of our bulkhead Atex fittings with these lamps with Manufacturer approval. They re-issued a blanket Certificate of Conformity.

These Discharge Lamps come in two types namely 'Integrated' and 'Non-Integrated'. The Integrated lamp is as per the diagram above, is where there is a glass section, ballast and cap. The Non-Integrated comes with just the glass shaped tube which is changed and the ballast etc. always remains inside the fitting.

The glasses come in many shapes, I have shown one shape in the diagram above. The ballast in this case is electronic and controls the voltage required for the lamp to start initially, and the current when the arc strikes from end to end.

Just like a standard fluorescent tube, when power is put onto the lamp there is an electronic starter which initiates the ionisation on the tube electrodes and the Mercury vapour carries the electrons around the shaped glass, which gives off photons of ultraviolet light and turns to visible light as it bounces off the white Phosphor. Obviously these lamps, being based on fluorescent, **CANNOT** be used with a standard leading edge dimmer.

Unfortunately some energy is still wasted in heat rather than light, but it is an improvement on the Incandescent Lamps mentioned earlier. The power consumption is also less than the Incandescent Lamp, but more than the modern LED (Light Emitting Diode) Lamps! What consumers may not notice is that these lamps lose a large percentage of their light in the first year.

Lifespan is around 15,000 hours and the lifespan will shorten if switched on and off regularly. When switched on, the lamp will take several seconds to reach full brightness as the gas warms up. The manufacturer's pack will always give a value of wattage equivalent to an Incandescent Lamp. Sometimes when the lamp is coming to the end of its life there is a buzz.

Always check installation advice from the manufacturers i.e. what type of environment or fitting can the lamp be used in? Has it got to be mounted in a certain position i.e. vertical cap up or down or universal? (This could have an effect on the safety and life of the lamp). Also check that the UK voltage is compatible with the rating of the lamp.

There are no large coils involved here, such as a magnetic choke so it is doubtful if these lamps have Power Factor Correction Capacitors. It there were thousands of them, which there are, then the Power Factor might be affected slightly, in which case other means must be sought by the electricity provider to correct their Power Factor i.e. large capacitors if at all required?

Sum up Questions on Compact Fluorescent Lighting (CFL):

Q1 – Can I use a standard Leading Edge Dimmer on Compact Fluorescent Lamps (CFLs)?

A1 – NO: If we look at this with our electrical knowledge and the information previously presented, discharge lighting has no metal filaments like an Incandescent Lamp or Halogen Lamp, so a Leading Edge Dimmer here could be catastrophic or at least producing inefficient quality of light.

A Leading Edge Dimmer could cause the tube circuit to go into a vicious restarting phase. Again we may investigate altering the waveform using Pulse Width Modulation.

Q2 – What is the white powder on the inside of the Fluorescent Tube?

A2 – PHOSPHOR: This is a toxic white powder coating which gives off a desirable light when 'Photons' bounce off it. Without a Phosphor coating there would just be an ultraviolet glow in the tube.

Q3 – Can I throw Compact Fluorescent Lamps in the dustbin?

A3 – DEFINITELY NOT! As well as toxic Phosphor they also contain a very small amount of Mercury.

Q4 – Do CFL Lamps have a ballast?

A4 – YES: Although electronic and very small in the base.

Q5 – What is an Integrated CFL.

A5 – ALL IN ONE: The Integrated lamp is where there is a glass section, ballast and cap.

Q6 – What is a non-Integrated CFL?

A6 – TUBE ONLY: The non-Integrated comes with just the glass shaped tube which is changed and the ballast etc. always remains inside the fitting.

Q7 – Do CFL lamps waste any energy in heat?

A7 – YES: But not as much as an Incandescent Lamp.

Q8 – What is the lifespan of a CFL lamp?

A8 – AROUND 15000 HOURS: They have a very long life span.

Q9 – How do I know a CFL lamp is coming to the end of its life?

A9 – LOWER BRIGHTNESS: The lamp will become dimmer and you may hear a slight buzz. These lamps may just go out at the end of their life, unlike an Incandescent Lamp which usually blows on switch on.

Q10 – Why are CFL lamps dim when first switched on?

A10 – HEAT UP THE GAS: It takes a few seconds for the gas to heat up from cold and the ionisation of the Mercury to get to full light.

Q11 – Do they have to be mounted a certain way?

A11 – SOME OF THEM: Otherwise the heat produced can do damage. Always check the Manufacturer's instructions obtained with the lamp, or as explained on the packet.

Fluorescent Chokes/Ballasts:

Firstly, in a fluorescent fitting is it called a Ballast or is it called a Choke? Some people term them as the same thing as they both limit the flow of electric current, but they are very slightly different. Chokes are '**inductive**' by the mere fact that they have many turns of wire in them, similar to a transformer. Ballasts on the other hand, can also be **resistive as well as inductive.** The short answer as far as a fluorescent light is concerned is either will suffice. Chokes can also be electronic, but not the older type I am about to describe, and it is there firstly, to provide the high voltage for the tube flashover and secondly, to limit the current when it does!

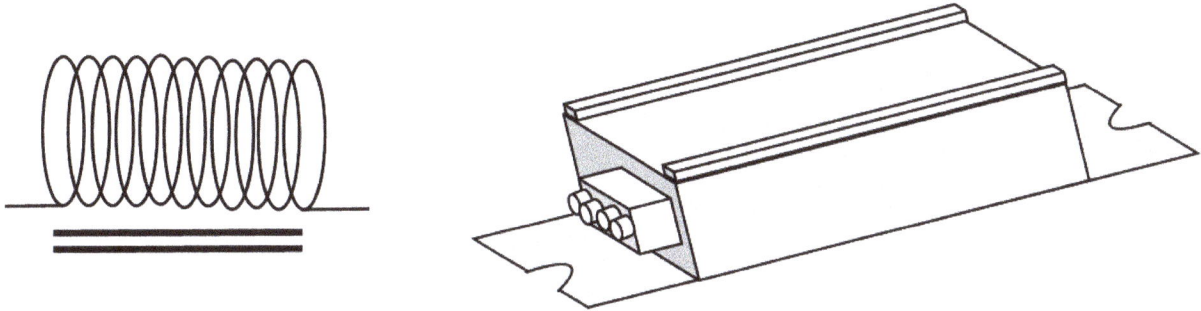

A Choke, as shown in the symbol and drawing above, is there to limit the current as once the arc strikes from one end of the tube to the other, it is more or less a short circuit. However, the current would be enormous and could in fact be so large that it would destroy the tube. So how does this unit work? Well, it limits the current by means of a magnetic field forming round the coils and being absorbed into its core, this is a phenomena called '**magnetostriction**'.

Remember, the Choke is a unit with many coils of thin wire wrapped around an iron core, which in the magnetic field, is expanding and contracting at 50Hz. It may cause a constant buzz, which may be amplified by the fitting being a tin box. Most of the time you cannot hear the buzz, as the light is above you on the ceiling. Have known chokes to get very hot and cause the fitting to catch fire the buzzing becomes very loud, sometimes giving the metal chamber of the fluorescent a sharp tap near the location of the Ballast will ease the buzzing rattling iron in the core. Sometimes if the fitting is buzzing loudly and continuous then it could be coming to the end of its life.

In older fluorescent lights, the Choke will be the coiled magnetic type, but in modern fittings this unit will be electronic which will not be noisy. Magnetic units can be changed for a modern electronic one, but manufacturers must be consulted if you wish to do this especially if the fitting is Atex Certified. The electronic Choke will boost the voltage slightly as well as controlling the current.

With magnetic Chokes, the output frequency does not differ from the main supply, which in the UK is 50Hz, but electronic ones have a high output frequency, 10,000-20,000Hz, so things like lamp flicker will be removed. Starters, of course, are not required with electronic Chokes. There are two methods of starting with electronic Chokes: **Rapid Start and Instant Start.**

Rapid Start involves a small voltage from a separate winding being applied to the tube electrodes on switch on. Just prior to the tube flashover, as a preheat voltage, a larger voltage around 500Volts, called a '**starting' voltage,** is applied to initiate the arc. This process will cause a slight delay of around 1 second. The tube may glow first at the ends. The preheat voltage means that the strike voltage need not be so large. No separate starters here of course!

Instant Start is very different from Rapid Start. Here a high voltage, around 600Volts, is applied straight away without the preheat voltage. No separate starters here either. This may be called a '**Cold Cathode**' start because there is no pre-heat. This is likely in an Atex Exe fitting and in older certified fittings where the tube may only have one much thicker pin (Mono-pin) and not two (Bi-pin).

Sum up Questions on Chokes & Ballasts:

Q1 – How does a magnetic choke limit the current in a fluorescent tube?

A1 – MAGNETIC FIELD: A Magnetic Choke limits the current by the magnetic field forming around the coils and being absorbed into its core, this is a phenomena called **'Magnetostriction'**. Chokes these days can be electronic.

Q2 – Is a Choke the same as a Ballast?

A2 – NOT QUITE: They are very slightly different, but they both achieve the same objective. Chokes are **'inductive'** by the mere fact that they have many turns of wire in them similar to a transformer. Ballasts on the other hand can also be **resistive as well as inductive**.

Q3 – If I change the fluorescent tube for an LED tube is the choke/ballast required?

A3 – YES and NO: You would with a type 'A' tube, but not with the type 'B' tube. The wiring should be modified to not include the Choke/Ballast. Manufacturer will advise.

Q4 – Why are Magnetic Chokes/Ballasts being replaced by electronic ones?

A4 – WEIGHT: The fluorescent fittings would be much lighter if electronic Chokes/Ballasts were used especially if it was a double fluorescent fitting.

Q5 – Why do some fluorescent fittings 'buzz'?

A5 – ALL FLUORESCENTS BUZZ! This only occurs in fluorescent fittings with magnetic Chokes/Ballasts. In most fittings the 'Buzz' is very quiet and barely audible. The sound comes from the Choke/Ballast and it is the magnetic field constantly expanding and collapsing on the laminated core to control the current. Remember the fitting is just a metal box which may amplify the sound.

Q6 – Can I change the magnetic Choke/Ballast for an electronic one?

A6 – YES: It is a simple task for an Electrical Technician to replace the magnetic Choke/Ballast and, as above, it will make the fitting much lighter. Remember, if the fluorescent fitting is certified i.e., Exd – Flameproof, Exe – Increased Safety or Exn – Reduced Risk, **the manufacturers must be consulted before this operation is completed.**

Q7 – If the fluorescent fitting has a starter is this required with the electronic Choke/Ballast?

A7 – NO: The electronic unit will incorporate the starting method.

Q8 – What is a 'rapid start' Choke/Ballast?

A8 – A STANDARD FLUORESCENT CHOKE/BALLAST: This is the normal Choke/Ballast for a standard fluorescent fitting which involves a small voltage, from a separate winding, being applied to the tube electrodes on switch on, just prior to the tube flashover, as a preheat voltage and then at the same time, a larger voltage around 500Volts, called a **'starting voltage'**, is applied to initiate the arc.

Q9 – What is an 'instant start' Choke/Ballast?

A9 – COLD CATHODE: No preheat. Here a high voltage, around 600Volts, is applied straight away without the preheat voltage. In older **certified** fittings the tube may only have one much thicker pin called a 'Mono pin' and not two, as in the standard Bi pin.

Induction Lamps:

You may not have heard of these induction lamps before, but they are used in abundance in certain situations such as high bays which may be hard to reach. Their life span is phenomenal and I mean years. I will let you read the description below and then make up your mind. Again, this is called a discharge lamp as there are no glowing elements.

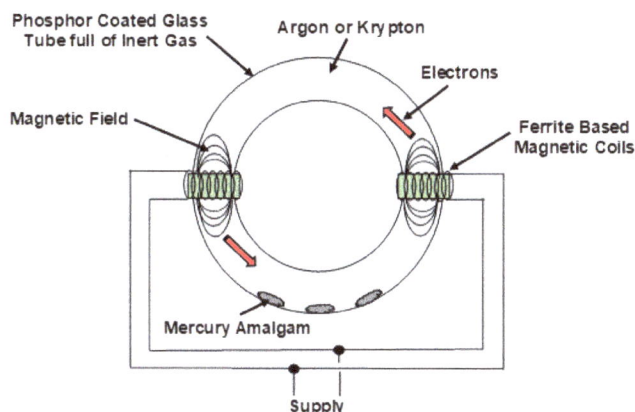

External Inductor Lamps are very different from other forms of lighting that we have talked about. They come in all sorts of shapes and sizes and some even resemble a traditional light bulb as in the diagram below, with screw or bayonet cap bases.

These traditional shaped lamps may have internal magnets which make for a smaller more compact shaped lamp.

The lamp on the left is shaped as a traditional round fluorescent the one below as a standard lamp shape.

We tried one as an experiment in one of our stores. We found that the lamp still caused some radio interference despite the ferrite coils otherwise, it was the same as a circular fluorescent, and dimmable, where a fluorescent would not be. It lasted years!

In the type like the round tube in the above diagram, or the diagram to the right, which is in the shape of a standard lamp, we have a glass tube full of inert gas.

On the round tube on either side on the outside of the tube is a ferrite based magnetic coil. When energised with a high frequency supply, it causes a magnetic field to form around them and 'excited' electrons to be emitted into the tube without any physical filaments.

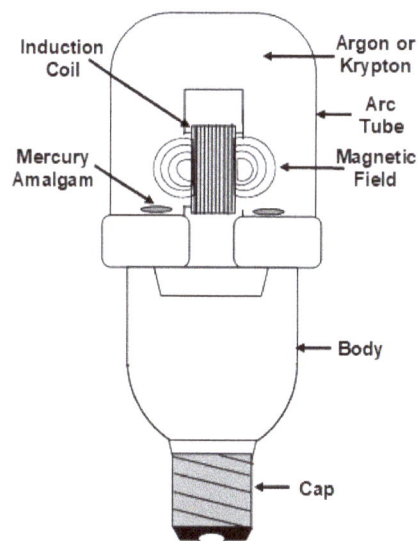

On the internal Inductor Lamp in the diagram to the right there is only one ferrite induction coil and you can also see in the diagram how the magnetic field is formed. The inside of the tube in both cases is coated with phosphor to give a warm visible light. Again this shape lamp will be installed in hard-to-reach locations and will last several years.

This process is called '**electromagnetic induction**'. The electrons in turn, collide with and vaporise the Mercury Amalgam and ultraviolet radiation is formed. This bounces off the internal phosphor coating to produce visible light, a process called fluorescence. What we are actually looking at is a fluorescent lighting effect without any physical electrodes, so the life expectancy of this lamp is considerably more than any other - we may be looking at many thousands of hours. (**Mercury amalgam is Mercury chemically combined with other metals.**)

These discharge lights would be ideal for places which are hard to reach such as high bay lighting in a 24/7 busy location like a warehouse or car parking lots which may be difficult to reach and where isolation of the lighting system may cause severe disruption due to dark areas. They can be obtained in the form similar to the diagram above or as a lamp similar to a compact fluorescent with Inductor Coils outside or inside. A Ballast would be required.

As far as lighting goes these may be on a par with LEDs but I don't actually know which would be the most expensive. but remember that these lamps will last for several years even if they are left on 24-7.

Sum up Questions on Induction Lamps:

Q1 – Does an Induction Lamp have filaments?

A1 – NO: It consists of a glass circular sealed tube full of inert gas with two induction coils, opposite each other around the glass. These coils cause an electromagnetic induction field which cause Mercury Amalgam deposits within the tube. Light is created by electrons colliding with the vaporised Mercury Amalgam. The inside of the tube is coated with Phosphor.

Q2 – Like a standard fluorescent does this lamp have a Ballast?

A2 – YES: The Ballast creates the high frequency required to make the lamp work. You could call this unit a frequency generator in this case rather than a Ballast. It usually rests on the reliability of the Ballast to establish the life span.

Q3 – Can I use a dimmer on these lamps?

A3 – YES.

Q4 – Does the lamp take time to warm up?

A4 – SLIGHTLY: Almost straight from switch on, but full brightness is achieved in the time it takes for the Mercury Amalgam to vaporise.

Q5 – What is the life span?

A5 – YEARS: These lamps have a long life, many thousands of hours. Sometimes the Ballast will fail before the lamp. With the lamp itself you could be looking at 100,000 hours!

Q6 – Can these lamps be obtained in a bulb shape?

A6 – YES: It does not have to be a circular tube.

Q7 – Can the Induction Coils cause radio interference?

A7 – YES: This is a problem. If there are many of them, radio interference could be a drawback. Some instrumentation may also be affected. Manufacturers will give advice.

Q8 – Are these lamps expensive.

A8 – YES: They are at the moment, which is a drawback, but you should consider the life span to see the value.

Q9 – Is heat produced by the Induction Coils?

A9 – YES: As far as the larger ring lamps are concerned, the design of the lamp allows for easy escape of heat. It is the smaller bulb type induction lamps that have the heat problems as there may be little escape for the heat.

Q10 – What is the efficacy of these lamps? (Watts in ratio to light out)

A10 – UP TO 80 LUMENS PER WATT: This is compared with LED at 130+ Lumens per Watt. The advantage over LEDs may be their life span.

Q11 – Can I install Inductor Lamps in Atex light fittings?

A11 – NO: Not without Manufacturer approval. Heat dissipation would have to be looked into very carefully!

Ultraviolet Black Lamps:

The diagram below shows the light spectrum visible light - Red, Orange, Yellow, Green, Blue, Indigo and Violet. This is what is emitted from most lamps, assisted by the white phosphor coating on the inside of the glass in some cases.

Just out of interest, can you remember the saying - **R**ichard **O**f **Y**ork **G**ave **B**attle **I**n **V**ain to remember the colours in the rainbow? This visible spectrum is formed by the shape of the sun bouncing off the **front** of the water droplets in the atmosphere, usually when there is rain and sunlight together. Another point of interest is that rainbows are a complete circle (the shape of the sun) and we only see half of it on earth. So, looking at the rainbow spectrum, this is light that we can see.

Apart from the lamps or tubes actually being black or dark blue, the actual emitted light is called 'Black Light', which is outside of the human visible light spectrum. So on the red spectrum this light is called infrared and on the violet spectrum it is called ultraviolet. Beyond these are X rays, gamma rays, alpha waves, beta waves, radio waves etc.

So, if we take a lamp and instead of a White Phosphor coating or clear glass and we use the black **'Wood's Glass'** we can turn it into a Black Lamp. This could be a Mercury Vapour, Metal Halide Elliptical as above, a Fluorescent Tube, Incandescent Lamp or a Compact Fluorescent **(CFL)**

A 'Wood's Black' Lamp, named after Robert Williams Wood, as the name suggests is a lamp where the bulb is made of 'Wood's Glass' and is totally black. This lamp gives off pure **long wave** Ultraviolet radiated light which, without the coating of **White** Phosphor is not visible to the eye. If, for instance, if you take a fluorescent tube and remove the White Phosphor coating and replace it with the 'Wood's' black filter coating from the inside, the tube would just be an ultraviolet glow. Some of these lamps have a special filter inside of the lamp blocking visible light and these are called **'Black Light Blue' (BLB)**.

Some Black lamps do not have this filter and these lamps can be elliptical as per the diagram above or many other types for example fluorescent tubes. They are used for light effects in discos, makes anything white really stand out, insect traps and in a whole range of medical uses. These lamps are designated **'Black Lamps' (BL)**. The light may be harmful to the human eyes depending upon type.

Although the light is not visible as, say a Mercury Blended Lamp, it could hurt your eyes to look at this lamp when it was working, and could damage eyesight if looked at for any length of time. Ultraviolet, along with infrared which is another non visible light, is in the light emitted from the sun, and it's the Ultraviolet can cause sunburn or browning of the skin. This being a type of radiation, it can cause burning of the skin and with long exposure can cause cancer..

We installed one of these lamps in a Certified ExnR Restricted Breathing fitting, to carry out an experiment on an instrument sensitive to Ultraviolet and although there was no visible light the fitting case got very hot - too hot to handle.

Sum up Questions on Black Lamps:

Q1 – Does a Black Lamp emit light?

A1 – **NOT IN THE VISIBLE SPECTRUM:** Only Ultraviolet light outside of human vision is allowed through the glass, so the lamp may appear not to be on. Sometimes, though, the lamp will have a violet glow.

Q2 – What type of glass are these lamps made of?

A2 – **WOOD'S GLASS:** Composed of Barium/Sodium/Silicate/Nickel. Invented in 1903 by American Physicist Robert William Wood (1868 – 1955).

Q3 – Are there different Black Lamps?

A3 – **YES:** There are Incandescent (has a metal filament), LED, Fluorescent, Mercury Vapour & Lasers. Incandescent black lamps use a filter to filter out everything, such as visible light, except longer wave UV-A which is required. Fluorescent tubes have an internal coating to filter out dangerous short wave UV-B & UV-C radiation and only emit longer wave UV-A less-harmful UV.

Q4 – Can these lamps damage eyesight etc?

A4 – **MOST DEFINITELY!** You would find them very hard to look directly at. The Wood's Glass UV is more damaging and has a shorter wavelength. The very short wave length UV-B & UV-C radiation, which would really do real harm, are filtered out. Ultra Violet (UV) is also known to damage skin with long exposure. Shorter waveforms are more dangerous because the light is more 'energetic'.

Q5 – As above can this light damage skin?

A5 – **YES:** It can cause skin ageing, wrinkles and destroy Vitamin 'A'.

UV is divided into A, B & C, UV-C is the most harmful as it has the shortest wavelength and is used mainly for sterilisation.

Q6 – Is normal Ultraviolet Light the same as Black Light Ultraviolet?

A6 – **NO:** The UV from Black Light (UV-A) has a shorter wave length than normal UV.

Q7 – Why do black items seem to have very stand out white marks in this light?

A7 – **PHOSPHORS:** Modern detergents and some materials contain Phosphors. Just like the white powder on the inside of a fluorescent tube, Phosphor turns UV into light so any Phosphors will stand out as light (bright white).

Q8 – What are Black Lamps used for?

A8 – **FORENSICS, ANTIQUE AND MONEY FORGERIES:** There are several uses. For example, fingerprint dusting powder stands out as well as body fluids in forensics and modern paints contain phosphors that were not used in the past.

Q9 – Can I obtain a Black Light Laser?

A9 – **YES AND TAKE GREAT CARE:** These lamps can be extremely hazardous to eyesight if not used properly and so correct eye PPE is required. The light emitted would actually be invisible and so outside of the colour spectrum (ROYGBIV)! One glance into the light with the naked eye will do permanent damage **IMMEDIATELY!**

Infrared Lamps:

We have already visited this diagram in the previous section on Ultraviolet Lamps. It shows the light spectrum of visible light - Red, Orange, Yellow, Green, Blue, Indigo and Violet which is emitted from most lamps, assisted by the white Phosphor coating on the inside of the glass in some cases.

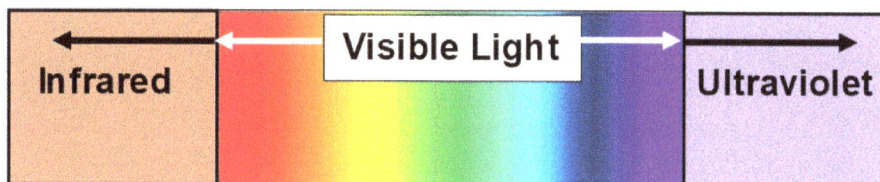

You will recall that these are the colours we see in rainbows, which is caused by the shape of the sun shining through and bouncing off the **front** of water droplets in the atmosphere usually when there is rain and sunlight together. So this is the spectrum we can see, but there is light we cannot see.

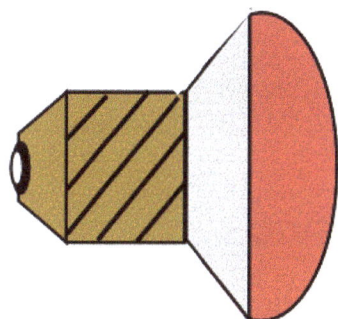

Outside of the visible light spectrum on the red side is Infrared and on the violet side Ultraviolet. Beyond these are X rays, gamma rays, alpha waves, beta waves, radio waves etc. The diagram on the left is of a lamp similar to the ones in health lamps used in physiotherapy for muscular pain; heat-lamps for reptile tanks or just for heating a specific area. Infra is the Latin word for 'below' so looking at the spectrum infrared (IR) is 'below' red. Ultra is 'beyond' so Ultraviolet (UV) is 'beyond' violet. I have owned one at home, but have not come across any on any plant.

In this case it does not matter if a large proportion of the Incandescent Lamp energy is given off in heat, because with electrotherapy this is what we are after; so Incandescent Lamps here would be suitable. With Incandescent Lamps, 90% of the lamps energy is given off in heat. The light from this lamp is a more focused beam using, say, a reflector, because we need it concentrating on a specific area. We might even add other metals to the Tungsten like **Chromium** or **Nickel** to alter the lamp's Infrared wavelength and make it more heat efficient.

Infrared Light can be split into three sections - **near, medium and far**. So **near** Infrared is closest to the visible light part of the spectrum and **far** is the furthest from visible light. In industry we use a thermal image camera to view hot objects, because it picks up the infrared emissions that our eyes cannot see. The army and police use infrared night vision to pick up objects including humans which give off heat in the dark. Many industrial gas detectors also use infrared to detect the amount of gas.

We used a thermal image camera to take an image of a large motor control centre by walking along viewing the starters, isolators and circuit breakers, to see if any of them were getting unusually hot. We saved many burnouts using this procedure.

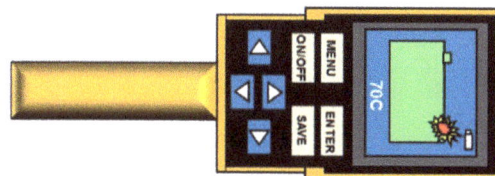

Some people ask, will Infrared pass through glass? The answer is, some of it and especially **near** Infrared, but quite a lot of energy from the **medium** and **far** Infrared may be absorbed by the atoms and electrons in the glass.

Can Infrared burn me? The simple answer is **YES**. Infrared Medical Lamps heat up the skin to encourage blood to flow more into specific areas, along with white corpuscles to fight infection. However keeping the Infrared on too long can cause burns, blood pressure, blotching of the skin etc. One very important health issue to be aware of is that it can affect your sight if you look directly at the Infrared Lamp glass.

Sum up Questions on Infrared Lamps:

Q1 – Can I see Infrared Light?

A1 – **MAYBE:** Depending upon which science paper you read. Some say Black Light Ultraviolet & Infrared is invisible whilst others say it is visible. Infrared Light just has a longer wavelength than visible light.

Q2 – Does Infrared Light damage skin?

A2 – **NOT NORMALLY:** Infrared Light enhances the skin and muscles, but like everything in excess can do harm usually in the way of burns.

Q3 – Are there different types of Infrared Lamp?

A3 – **YES:** Usually these lamps are used for healing where heat is required. Incandescent Lamps can be used here because most of their energy is in heat. Other sources of Infrared Light may come from **L**ight **E**mitting **D**iodes.

Also the Infrared Light itself can differ and be obtained as **Near** (1-5 microns), **Medium** (5-30 microns) and **Far** (30-300 microns), and what is called the '**submillimetre**' (300 microns to 1mm). **Near** being the closest to visible light and **Far** being furthest away. A remote control would be **Near**.

Q4 – Are the filaments of Infrared Lamps Tungsten?

A4 – **YES:** But they may add Chromium or Nickel to alter the wavelength and make the lamp more efficient.

Q5 – Are there other uses of Infrared Light besides medical?

A5 – **YES MANY!** Some examples being; Night-vision devices, thermal image cameras, TV remote controls, computer communication, telescopes for astronomy.

Q6 – Will Infrared pass through glass?

A6 – **SOME OF IT:** Especially if it is in the '**near**' band.

Q7 – Is Infrared Heat the same as Infrared Radiation?

A7 – **YES THEY ARE THE SAME THING:** Just a different title.

Q8 – Can Infrared light travel through smoke?

A8 – **YES:** Smoke and fog.

Q9 – Does the human body give off Infrared radiation?

A9 – **YES:** Which is why humans can be spotted by, say, a police helicopter using heat seeking device.

Q10 – Who discovered Infrared light?

A10 – **WILLIAM HERSCHEL:** In the 19th Century, Herschel, an astronomer, pioneered the use of astronomical spectrophotometry. In the course of these investigations, he discovered Infrared radiation.

Q11 – What is the speed of Infrared Radiation?

A11 – **APPROXIMATELY 186,000 MILES/SECOND!**

Sulphur Lamps:

This method of lighting is a totally new way of looking at light output. What we do here is 'excite' Atoms of Sulphur by using a Magnetron to generate microwaves, as it would do in a microwave oven. Let us have a look at how Sulphur Atoms are excited artificially into a type of plasma.

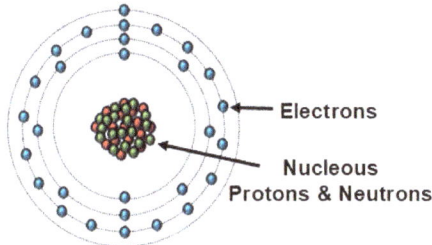

Left is an Atom in its normal state with Protons and Neutrons in the Nucleus and Electrons in fixed orbits. The Electrons closest to the Nucleus are low energy and the ones further away are higher energy. If we put a certain power supply to the Atom we can encourage the Electrons closest to the Nucleus to 'jump' into higher energy orbits and this is called 'exciting the Atom'. As we do with a Laser.

Some Electrons in the higher orbits can move down towards the Nucleus to take their place and this movement releases, in this case, an extremely bright light. This method of lighting will give off the same light as hundreds of standard light bulbs which, of course, on its own is far too bright to be practical.

We start off the basic design with a revolving Quartz sphere full of Argon gas that also contains a small amount of Sulphur. A Magnetron is used to bombard the sphere with microwaves at a certain frequency. The microwaves excite, as mentioned above, the Sulphur atoms into a Plasma. The lamp was designed to rotate and a cooling system set up to cool the bulb from the extreme heat of the Plasma. The surrounding mesh is for protection from the dangerous microwaves.

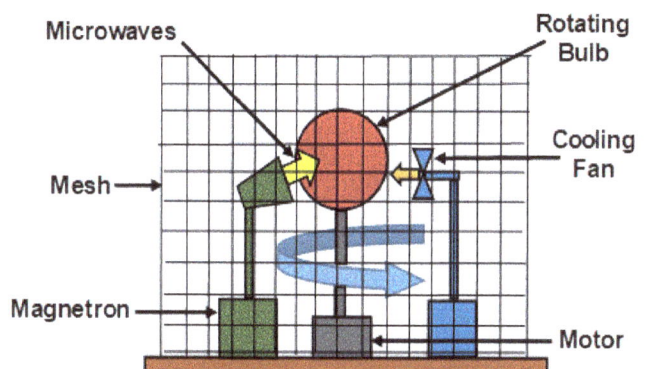

If we take our Sulphur Lamp and use it as the diagram above, as already mentioned, it would be far too bright to use on its own, so the next diagram and explanation further demonstrates how it is used.

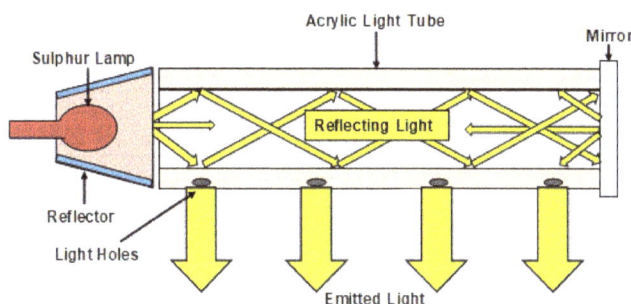

Looking at the diagram, left, the Sulphur Lamp is inserted into a conical reflector so that the light is reflected down an 'Acrylic Tube' with a prismatic reflective film. The tube has a mirror at the opposite end to the lamp so that all the light is reflected back into the tube. As above an external cooling system must be in place to cool the lamp.

As you can see in the diagram above, the prismatic walls of the light tube have a series of holes. I have only shown four but in fact there would be many more, to allow the light inside of the tube to escape to the outside. The life of this lamp could be as much as 100,000 hours as there are no solid electrodes or filaments to wear out. Any such material as Tungsten, say, for filaments would be quickly attacked by the Sulphur.

I have seen one of these lamps when a Rep demonstrated these lamps for us and I took quite a few notes and diagrams at the time. The amount of equipment required is huge i.e., Magnetrons, mesh, cooling fans, rotating motors, light tubes not to mention the lamp itself: so the cost would be huge. Potential users would have to weigh this up against the amount of other types of lamps required to take its place. I could not even begin to work out the efficacy.

Sum up Questions on Sulphur Plasma Lamps:

Q1 – What is an 'Excited' atom?

A1 – **WHERE ELECTRONS JUMP ORBITS:** By putting a certain power onto the atom, we can cause Electrons closest to the nucleus to jump into a higher orbit. Other Electrons can move down to take their place.

Q2 – Are the industrial Sulphur Lamps dangerous?

A2 – **NO:** However is can be if you expose the Magnetron during maintenance with the lamp switched on. It will cause microwaves to form, the same as a microwave oven. Also Magnetrons contain a substance called 'Beryllium Oxide' and 'Thorium' both of which are **VERY** damaging to health, if breathed in.

Q3 – What is Plasma?

A3 – **A TYPE OF SUPERHEATED GASEOUS SOUP:** Electrons are pulled out of their orbits by a device called a Magnetron and form this medium of charged Ions & Electrons. Plasma is said to be the fourth state of matter. Plasma is of course electrically conductive. A Sulphur Lamp turns the Sulphur in its centre orb into Plasma.

Q4 – Are Sulphur Lamps efficient as to brightness?

A4 – **TOO EFFICIENT:** If the light was left as it exited the lamp it would be far too bright to be used as lighting, so it has to be put through an acrylic lighting tube to reflect the light back into itself, leaving a nominal light output to do its job. They can also be used for indirect lighting.

Q5 – Have Sulphur Lamps got a good lifespan?

A5 – **YES!** Up to 50,000-100,000 hours - the efficacy is enormous - wattage in to light out ratio. Other equipment such as Magnetrons, motors and cooling fans may fail first!

Q6 – Are these lamps expensive?

A6 – **YES:** Due to the amount of auxiliary equipment that has to be used i.e., Magnetrons, motors, cooling fans etc.

Q7 – Can these lamps cause radio interference?

A7 – **YES:** It is the Magnetron that causes the problem.

Q8 – Why not use a Filament?

A8 – **SULPHUR:** It would attack any Tungsten or in fact any metal filaments!

Q9 – Are these lamps environmental friendly?

A9 – **YES:** Sulphur does not pose a threat to the environment, and this type of lamp does not contain any Mercury.

Q10 – How are humans protected from the hazardous microwaves?

A10 – **A PROTECTIVE MESH:** This is put around the light sphere which is bombarded by microwaves.

Q11 – What is the purpose of the electric motor?

A11 – **COOLING:** The motor spins the light ball so that the cooling fan can cool it.

Lasers:

These days Lasers are used extensively for a whole range of medical tools as well as industrial instrumentation and cutting tools We have all seen Laser light from Laser pointers and Laser light shows at fairs and big events, but these are nowhere near the Laser intensity required for cutting although they can still harm vision if shone into eyes.

So how do we channel light into such a concentrated beam that we could transmit great distances?

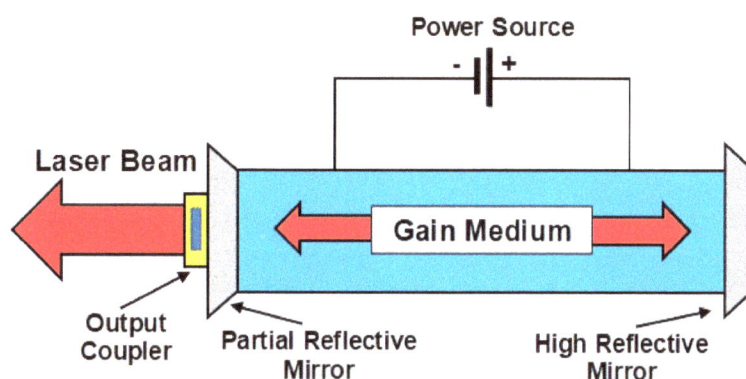

Well firstly 'Laser' stands for **L**ight **A**mplification by **S**timulated **E**mission of **R**adiation. We start off with a **'Gain Medium'** sometimes called an **'Active' Laser Medium** in a tubular vessel. As you can see from the diagram above, we have a vessel containing the Gain Medium, A Power Source and two mirrors, one very high reflective and one partially reflective mirror called an output coupler. This uses the Gain Medium to concentrate and amplify the beam.

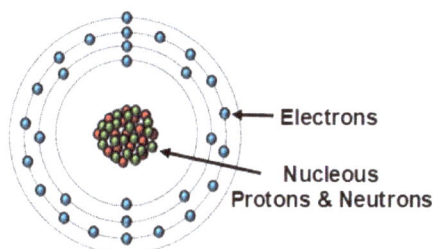

Left is an atom which, at the moment, is in its normal state with Protons and Neutrons in the Nucleus and Electrons in fixed orbits. The Electrons closest to the Nucleus are low energy and the ones further away are higher energy. If we put a certain power supply to the Atom we can encourage the Electrons closest to the Nucleus to 'jump' into higher energy orbits and this is called **'exciting'** the Atom!

Some Electrons in the higher orbits can move down towards the Nucleus to take their place and this movement releases another type of energy called a Photon. All this reaction is taking place inside of the Gain Medium which is inside our vessel with a mirror on either end. As these Photons are released they travel around the vessel, striking the mirrors, especially the high reflective one and are reflected back into the Gain Medium, colliding with other **'Excited Atoms'**. The whole process starts to amplify enormously, considering the millions of atoms and hence Photons in the Medium. This process is called **'Stimulated Emission'**. Gain Media can be gas or solid, ceramic or crystal.

So as Photons are released, caused by the excitation of the Atoms, many bounce off the high reflective mirror back into the medium. The energy source must keep supplying the Gain Medium with energy to enable the process to continue and this is called **'Population Inversion'**. Millions of Photons now escape the **Gain Medium** through the partially reflective mirror and the **Output Coupler**, which may contain a fibre optic directional guide and this is our Laser Light.

So how bright is a Laser and how far will the light travel? Well you must not point one at your or anyone else's eyes, as the light at close range could be many times brighter than the sun. As to, how far will the beam travel? Well, the answer would be several miles depending upon the size and type.

I am sure that Laser Technology is more complex than I have made it in this explanation, but at least you have an idea of the basics.

Sum up Questions on Lasers:

Q1 – What does Laser stand for?

A1 – **L**ight **A**mplification by **S**timulated **E**mission of **R**adiation!

Q2 – What is an 'Excited' Atom?

A2 –ELECTRONS JUMPING ORBITS: By putting a certain power onto the Atom we can cause Electrons closest to the nucleus to jump into a higher orbit. Other electrons can move down to take their place. (Excited Atoms)

Q3 – What is a Photon?

A3 – A PARTICLE OF LIGHT: We call this a bundle of **Electromagnetic Energy**. It's very dependent on frequency for its energy.

Q4 – How far can Laser light travel?

A4 – MILES: A laser pointer, for example, could probably be seen 10 miles+ away. Some idiots have shone laser pointers up at aeroplanes and the pilot's eyesight has been affected.

Q5 – Are there different types of Laser?

A5 – YES: The difference is the type of 'Gain' Medium gas/material. Gases include: Helium, Argon, Neon. Crystal: Ruby. Chemicals: Various. All to do with atoms being exited (Q2 above)

Q6 – Are Lasers hazardous to health?

A6 – MOST DEFINITELY: They can seriously damage eyesight if shone in eyes. A Laser scalpel can cut through tissue etc.

Q7 – What is **'Population Inversion?'**

A7 – THE ACTION OF EXCITING THE MOST ATOMS

Q8 – What is a Maser?

A8 – **M**icrowave **A**mplification by **S**timulation **E**mission of **R**adiation: Excited atoms can amplify radiation. A Laser is in fact a type of Maser!

Q9 – Can I cut steel with a Laser?

A9 – YES: Steel sheet up to 3cm would be no problem!

Q10 – What is a **Laser Gain Medium?**

A10 – A MATERIAL THAT AMPLIFIES LIGHT: Sometimes called an **Active Laser Medium.** Excited atoms and Photons keep moving back and forth in the Gain Medium amplifying all of the time and the final result is Laser light.

Q11 – Does Laser light exist naturally?

A11 – NO: Laser light is created artificially.

Q12 – Are there laser weapons?

A12 – YES - Of Sorts: The Russians have a crude laser weapon called 'Peresevet'

Plasma Lighting:

Plasma is the 4th state of matter. When people talk about plasma lamps they immediately think of the novelty lamps that were around in the 1980s and looked something like the diagram below:

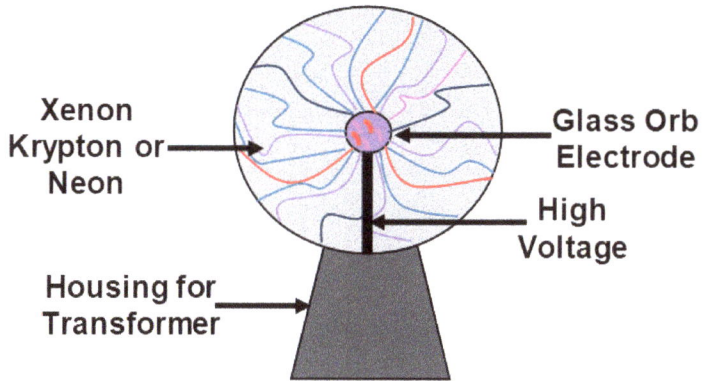

Xenon Krypton or Neon

Glass Orb Electrode

High Voltage

Housing for Transformer

If you can remember these fun lamps where the arcs inside would follow your hand wherever you placed it on the glass orb, but they will never shock you. They did not actually emit much light and were purely for fun. We can thank Nikola Tesla for the idea although it may have been invented by accident!

The outer glass sphere has more electrons than protons so is negatively charged.

Several thousand volts at around 30 Kilohertz is fed from a step up transformer into the 'Orb' electrode containing a Tesla coil. The Tesla coil vibrations are how the high frequency is formed, ionising the gas and causing the multi-coloured streamers, harmless inside the sphere.

By using the idea of the novelty lamp, we can now enhance it to make useful lighting units. First, we need a power supply feeding a Magnetron, a device for emitting short radio waves, not too dissimilar from your microwave oven. These microwaves require our next device, which is a wave guide to direct the microwaves into the lamp section.

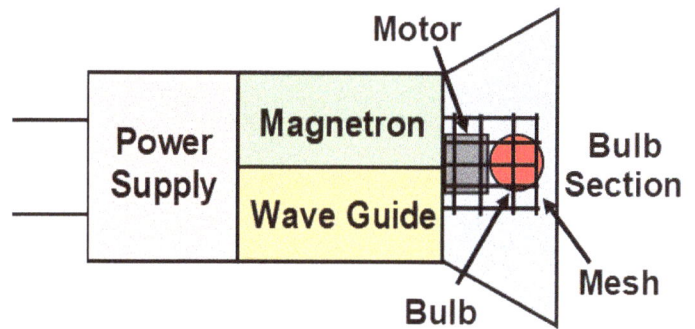

Motor

Power Supply

Magnetron

Wave Guide

Bulb Section

Mesh

Bulb

These may be called LEP lamps (**Low Energy Plasma**) or later version HEP lamps (**High Energy Plasma**)

As we discussed in the Sulphur Lamp section, a mesh will stop the dangerous microwaves from escaping. So here around the inside of the bulb section is a mesh. As the diagram at the top shows there is an **'orb'** in the lamps it is the **'bulb'** which emits the light.

Inside of the 'bulb' is a small amount of noble gas or Halides, Sodium, Sulphur or Mercury which makes it similar to a Sulphur lamp. The bulb is 'excited' which we mentioned in the Laser section, where we cause electrons in lower orbits to 'jump' into a higher orbit. The result being Plasma, Photons and very bright light.

Plasma is like a very hot gaseous soup, it' the best description that I can come up with. Like the description earlier on the Sulphur Lamp, the bulb has a small motor type unit which rotates it at a slow, constant speed so that the light and plasma is evenly distributed. These lamps are very expensive, but have a very high luminaire efficiency, which allows them to be the nearest to natural daylight as is possible.

One of the problems that may be encountered is that the Magnetron, not the lamp, has a finite lifespan of around 3,000 hours and are quite expensive to replace. If you do have to replace this device **BE VERY CAREFUL DO NOT** operate the Magnetron outside of its surroundings in the lamp as microwaves are very dangerous to humans especially in the direction of the eyes.

Never try to strip down a Magnetron, as inside they contain a substance called **Beryllium Oxide and Thorium (minutely radioactive)** which can be **SEVERLEY** unhealthy, and in some cases fatal, if breathed in.

Sum up Questions on Plasma Lamps:

Q1 – Are Plasma lamps and Sulphur Lamps the same?

A1 – **NOT QUITE:** Sulphur plasma lights came first. They both produce Plasma in an orb but there are slight differences. For instance, in this case the 'orb' is filled with noble gas e.g. Halides, Sodium, Sulphur or Mercury.

Q2 – What is an 'Excited' Atom?

A2 – **CAUSING ELECTRONS TO JUMP ORBITS:** By putting a certain power onto the atom we can cause electrons closest to the nucleus to jump into a higher orbit. Other electrons can move down to take their place. (Excited Atoms)

Q3 – Are the industrial Plasma Lamps dangerous?

A3 – **NO:** In general use they are safe, however do not expose the Magnetron during maintenance with the lamp switched on! Microwaves will form just the same as a microwave oven.

Also Magnetrons contain a substance called Beryllium Oxide and another called Thorium (slightly radioactive) which is **VERY** health damaging if breathed in.

Q4 – What is Plasma?

A4 – **A TYPE OF SUPERHEATED GASEOUS SOUP:** Electrons are pulled out of their orbits by a device called a Magnetron and form this medium of charged Ions & Electrons, which is said to be the fourth state of matter. Plasma is of course electrically conductive.

Q5 – Are there different types of plasma lamps?

A5 – **YES:** There are LEP (**Low Energy Plasma**) and HEP (**High Energy Plasma**)

Q6 – How are humans protected from the hazardous microwaves?

A6 – **A PROTECTIVE MESH:** A mesh is put around the light sphere which is bombarded by microwaves.

Q7 – Are these lamps expensive?

A7 – **YES:** Because of the amount of auxiliary equipment that has to be used i.e. Magnetrons, motors etc. Later lamps were designed with materials which did not have the heat build-up, removing the need to revolve the orb. Another problem is that they are big and bulky.

Q8 – What is the purpose of the electric motor?

A8 – **EVEN DISTRIBUTION** - Of light and Plasma.

Q9 – What is the lifespan of these lamps?

A9 – **THE PROBLEM IS THE MAGNETRON:** It has a lifespan of around 3,000 hours and Magnetrons are expensive to replace.

Q10 – What is the efficacy of these lamps?

A10 – **GOOD!** Efficacy is the ratio of the amount of power that goes in versus the amount of light out. Around 90 lumens/watt. The problem here is the lamp could not sustain its Plasma under 1000 Watts which makes it use more power.

Germicidal UV Lamps:

If we take a fluorescent tube with no Phosphor coating on the inside, a Quartz glass tube instead of glass and turn the power onto the electrodes, we would just have an ultraviolet glow and a very tiny amount of visible light (as below). We would still require a ballast as we would for a standard fluorescent tube, to control the current flow once the arc is struck. So without doing too much we have produced ultraviolet.

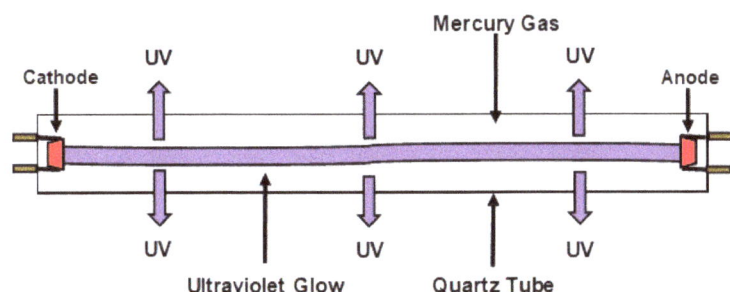

We can take this effect one step further with the design of a medical lamp using Ultraviolet Light to destroy bacteria and harmful pathogens that may develop and cause diseases in humans by altering their DNA. It is the wavelength of the UV that is important and has to be around 150 nanometres.

The beauty of this therapy is that it is non-invasive, simple and relatively inexpensive. As well as uses in medical environments, a UV (Ultraviolet) Lamp is used in other environments such as the food and water industry, where it also destroys micro-organisms. In this case we are not talking about a black lamp as described in an earlier section. This is a different form of UV.

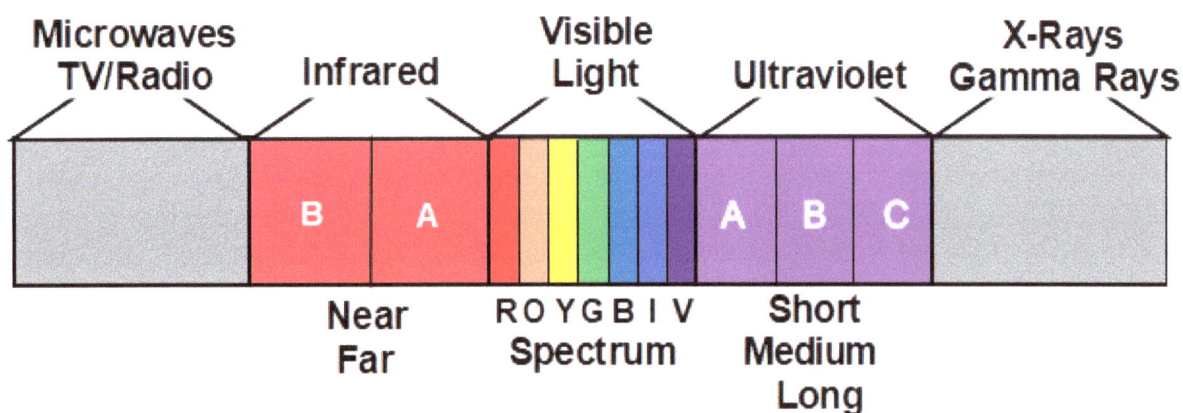

Looking at the spectrum above, on either side of the visible light section there is Infrared (left) and Ultraviolet (right). Infrared is divided into two - near (A) and far (B) - and ultraviolet divided into three - short (A), medium (B) and long (C) - so the correct value would have to be selected. It is the Ultraviolet at a certain value as above (UV-A, UV-B & UV-C) that we are interested here

The lamp in the diagram on the right is called a Corn Lamp (no points for guessing why!) These lamps are a type of health lamp, and they can be bought on the Internet for domestic sterilisation. An example of where they might be used is in bedrooms on beds to get rid of bedbugs etc. They can also be used in bathrooms, toilets & kitchens etc.

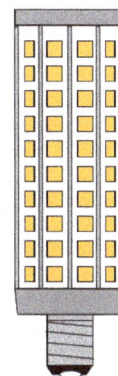

Any Germicidal Lamps can be extremely dangerous to humans if not used correctly and using equipment made for such lamps. This ultraviolet lamp, as for any UV, goes for the skin and eyes and with an intensity can cause effects similar to being in the sun for large amounts of time. Wherever the lamp is being used, especially for domestic use, it is very important that eyes and skin are not exposed to it and **that includes pets.** When the lamp is being used in either medical or domestic situations, PPE is recommended.

When the system is operating, strict controls must be in place to prevent people from walking in and out of a room without any PPE. Warning notices must be put onto doors etc. If in domestic circumstances, children must be prevented from coming into contact with these lamps.

Sum up Questions on Germicidal UV Lamps:

Q1 – Are there different types of Ultraviolet Light?

A1 – YES: There are three types - **UV-A, UV-B & UV-C.** These lamps use the UV-C 200-250nm wavelength.

Q2 – How does the lamp carry out its purpose?

A2 – STERILISATION/PURIFICATION: The lamp destroys the ability of bacteria, viruses & pathogens to multiply by altering their DNA. No micro-organism has been found to be immune. I have mentioned viruses, but they are a non-living entity so by using this method you simply de-activate them.

Q3 – Where would this lamp be used most?

A3 – WATER, FOOD & MEDICAL INSTRUMENTS: Kills microbes in seconds!

Q4 – Is this radiation, UV-C, a risk to humans & animal?

A4 – YES HIGHLY IN MOST CASES! UV-C, which germicidal lighting uses, is a risk to skin and eyes. Some damage may be immediate, but some may not manifest itself for several hours. (200-250nm)

Germicidal lamps which are perfectly safe for human use can be obtained, but always check the data! Also, lamps can be positioned so that there is no human exposure, say, above head height again check the data!

Q5 – Would UV-A & UV-B be effective as germicidal lighting?

A5 – NO: UV-A (around 400nm) would have no effect at all. UV-B (around 300nm), although may be effective, it poses a higher health risk as it can penetrate the skin much more deeply than the others. If any of the three UV radiations were going to give you skin cancer it would be UV-B.

Q6 – Can these lamps be bought domestically as well as for industrial use & laboratories?

A6 – YES: They can be used to sterilise rooms such as bedrooms (for bedbugs) and bathrooms. If used in bedrooms correct PPE should be used as protection against skin and eye damage. **REMEMBER THIS LAMP IS ALSO DANGEROUS AS REGARDS TO PETS!**

Q7 – Is this lighting more effective than liquid disinfectant?

A7 – NOT NECESSARILY: Liquid disinfectant deals as efficiently in some cases and sometimes it is a case of convenience.

Q8 – What is the typical lifespan of a germicidal lamp?

A8 – AROUND 2 YEARS.

Q9 – Does sunlight contain Germicidal UV Radiation?

A9 – YES AT TIMES: Depends upon which time of year and where on the Earth. Springtime and summer are peak times. However, nothing to cause the health problems with the UV-B & UV-C lamps.

Q10 – Can these lamps can also be obtained in LED form?

A10 – YES.

Carbon Arc Lamp:

The Carbon Arc Lamp is really a lamp of the past. They give quite a lot of light and would suit a large area such as a warehouse, or streetlights as they were used in the past. The huge problem with them is that the Carbon electrodes wear out quite quickly with the arc and have to be replaced on a relatively regular basis.

Around the middle 1800s, Jean Foucault invented a mechanical device that actually fed the Carbon electrodes together as the arc decayed them, which made the lamp last much longer without maintenance. Early lamps also emitted quite a lot of Ultra Violet which was a problem in itself, and there was no globe until later. The Carbon Arc Lamp was replaced by the Xenon Lamp.

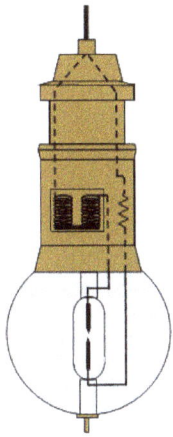

The diagram left shows an example of one type of Carbon Arc Lamp. It was designed by two inventers called Elihu Thompson and E. W. Rice and was called the **Thompson Arc Lamp**. This lamp contained a balancing coil and toroidal reactor in the top to control the arc and iron out the unstable electricity feed to the lamp. The balancing coil at the top could be huge - much larger than shown in the diagram on the left. The arc was in a pod in the centre of the globe.

As already mentioned the Carbon electrodes had a finite life and eventually required to be replaced. At the beginning of the 1900s Charles Steinmetz developed a Carbon arc lamp that replaced the pure Carbon electrodes with Iron Ore (**Magnetite**) Carbon and this modification gave the electrodes a much longer lifespan. The Carbon electrodes start off nearly touching and are forced apart by a process called Incandescence due to high resistance.

The Carbon Arc Lamp was replaced by the **Xenon High Pressure Lamp**. In this lamp there is a very high pressure pod full of Xenon, which contains the arc. High pressure makes the lamp more efficient, but unfortunately makes it very hazardous for the people who had to replace them. If dropped they could resemble a bomb with the high pressure causing a small explosion sending glass shrapnel towards anyone nearby.

Another type of Arc Lamp is a **Kleig Arc Lamp** used in film making in the 1930s because of its brightness. The lamp was invented by the brothers John & Anton Kleigl. They used a **'Fresnel lens'** which is a similar type to those used in a lighthouse to beam the light, with a very shiny back reflector.

Now we turn our attention to the **'Yablochkov Candle'** which was invented by Russian Electrical Engineer Pavel Yablochkov. The lamp is made up of two Carbon Rod electrodes separated by an insulator and in the day would most likely been made with plaster of Paris.

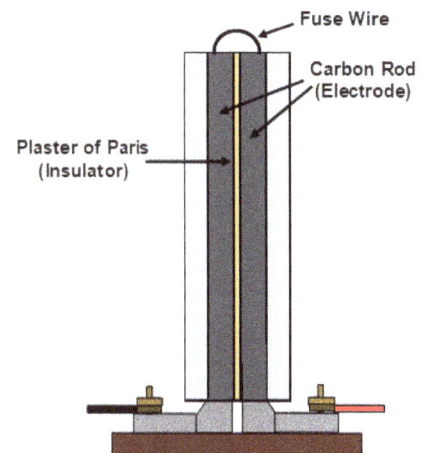

The idea was that the lamp (Candle) is switched on and the current melted the fuse wire at the top and so set an arc going between the two Carbon electrodes. This in turn caused the Carbon Rods to slowly burn down taking the insulator with them. Of course once the fuse wire had gone the arc cannot be re-activated although this was modified in later lamps. Another disadvantage was that the lamp only lasted several hours.

Now that we have looked at several Arc Lamps, and I am sure that there are more, I am satisfied that you have a good idea of how they worked and the advantages and disadvantages that inventors and users were faced with.

Sum up Questions on Carbon Arc Lamps:

Q1 – Were there different forms of this type of Arc Lamp?

A1 – YES: Carbon Arc DC, Carbon Arc AC, Flame Arc & Magnetic Arc. Carbon Arc described here was the most common.

Q2 – Are Carbon Arc Lamps used today?

A2 – NO: The Carbon electrodes have a finite life and burned away very quickly so had to be replaced frequently, so the lamp was not practical.

Carbon particles constantly flow from Positive to Negative. Because the particles leave the Positive it becomes hollow and because they collect on the Negative it becomes bulbous to fit the Positive hollow.

A near comparable process is DC Carbon Arc welding. The welding rod is always the negative so that Carbon particles do not flow into the weld. Graphite can be used instead of Carbon which has the same chemical symbol 'C'

Q3 – What replaced the Carbon Arc Lamp?

A3 – THE XENON HIGH PRESSURE LAMP: Although the problem with this lamp was that it was like carrying a bomb, because if dropped, the high pressure would send glass shrapnel in all directions frequently causing injury.

Q4 – When was the Carbon Arc Lamp invented?

A4 – 1808: See history section.

Q5 – What was Carbon lighting used for?

A5 – CINEMA, STREET LIGHTING AND EVEN LIGHTHOUSES: Because the lamps in the street lighting had such a finite life they employed a man, whose full-time job was to go around the streets replacing the lamps. The lighting was good because of the intense brightness of the Carbon arc.

These lamps also generated small amounts of UV in the bands of UV-A, UV-B & UV-C which of course with long exposure could affect eyesight.

Q6 – What is this type of arc called?

A6 – A VOLTAIC ARC.

Q7 – Would the arc voltage be AC or DC?

A7 – AROUND 50 VOLTS DC: They would not have had AC when the lamp was invented. Later, AC was used with a type of balancing coil and a toroidal reactor, a type of ballast, at around 55-110 Volts AC.

Q8 – What temperatures are produced by the arc?

A8 – AROUND 3,600°C: This would vaporise the Carbon.

Q9 – Would a gas have been produced by the Carbon Arc?

A9 – YES: Carbon Monoxide (CO) which again was hazardous to humans.

Water & Air Cooled Lighting:

Water Cooled Lighting:

If we talk about water cooled electrics, many people would think it would be impossible. 'How can I get an electric lighting system that was water cooled?' Well they do exist, and the concept is not new, in the form of what are called **'Grow-lights'**.

This lighting is largely used in large greenhouses where they grow huge amounts of plants 24/7. In the past Sodium and Metal Halide could be water cooled. These days the lighting will be LED and although they run much cooler than an Incandescent light, in large numbers they do produce large amounts of heat and the water vessel in the lamp is a heat sink.

You might ask why we do not use air to cool the light fittings? Well the answer is **'well we could'**, but water would do the job much more efficiently because of its density compared with air. Water is around 1000 times denser than air.

So if we pump water round the system and the water gets hot where does it go to be cooled? Well, it enters a cooling system/heat exchanger or chiller, to cool the warm/hot water down ready to be re pumped around the system. A chiller is expensive, or, as below, the water can be collected in a reservoir before being re-pumped, both of which act as heat exchangers. Advanced systems can use the heated water for other purposes in the growing system such as radiators in a heating system which would in itself be a heat exchanger.

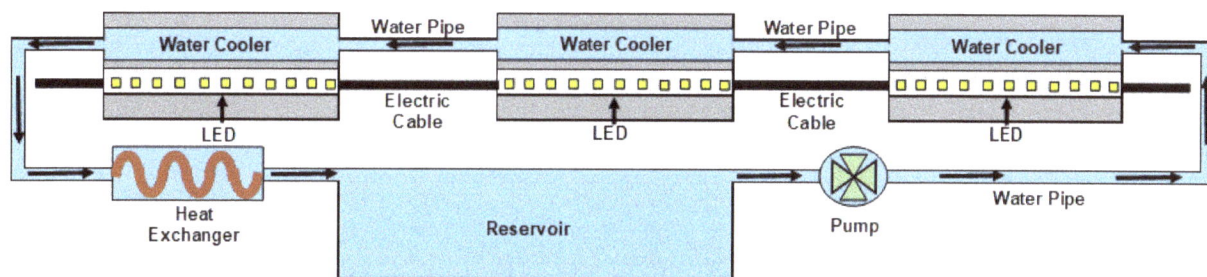

Air Cooled Lighting:

It is possible to get Air Cooled Lighting and again as the above liquid cooled lamps, we are talking mainly about **Grow Lights**. Here, cold air is forced over the reflector and lamp to provide the cooling. Light fittings can be bought with a built-in fan or you could build your own by buying a separate fan, although this may not be quite as efficient.

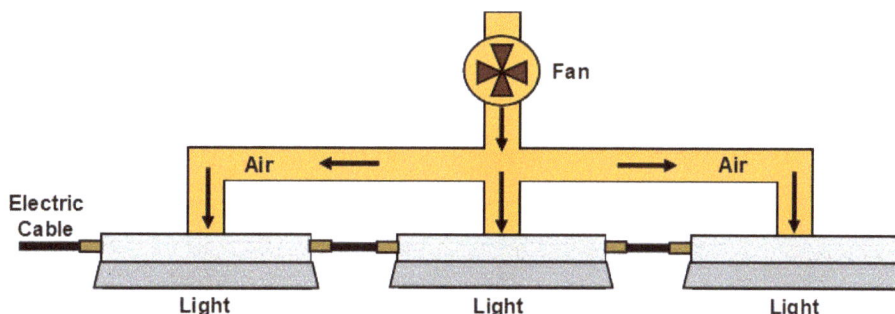

The diagram above, shows how the air system may be connected. High Pressure Sodium Lamps and Metal Halide Lamps can be air cooled, but these are slowly being replaced by LED lamps which although do not get as hot as the others, can produce heat if they are on 24/7.

Sum up Questions on Water and Air Cooled Lamps:

Q1 – Where does water fit into a lighting circuit?

A1 – IT DOES NOT: The water is used as a heat sink. The vessel of flowing water runs parallel with the lamp and absorbs the heat giving a more prolonged life expectancy.

They could, and do, use air but water is around 1000 times denser and more efficient so in the long run uses less energy for the same light. (Efficacy)

Q2 – Where is this type of system used?

A2 – GREENHOUSES: They are called **'Grow Lights'** where the lighting may be on 24/7.

Q3 – Are the lamps special?

A3 – NO: There are three common lamps: Fluorescent, High Intensity Discharge (Sodium or Metal Halide) & LED.

The High Intensity Discharge (HID) might require the most cooling. LED lamps will run the coolest.

Q4 – How does the water itself cool?

A4 – HEAT EXCHANGER AND/OR RESERVOIR: The water is constantly pumped around the system, sometimes through a heat sink into a reservoir where it is cooled.

A more efficient way but with an added cost, would be a chilling unit.

Q5 – Would this system be expensive?

A5 – YES VERY: All the equipment has to be bought and fitted initially. As well as the electric power for the lighting and the pump - if water, or fan - if air, requires constant power.

If a chiller unit is employed then this must be purchased, fitted, maintained and it requires constant power.

Q6 – Can the heat be used for anything else?

A6 – YES e.g. RADIATORS: Maybe in the offices instead of a heat sink in winter.

Q7 – Can the cooling system cope with a huge number of lights?

A7 – NOT USUALLY - DEPENDS WHAT TYPE OF SYSTEM: Usually around five lights per system, otherwise it could be inefficient. Manufacturers will give specifics.

Q8 – Can I use lamps that do not require a cooling system?

A8 – YES: But for growing plants, the type of light not be as efficient.

Q9 – What coloured lighting would Grow Lighting have to be?

A9 – PURPLE/BLUE OR ORANGE/RED: Plants reflect greens & yellows.

Q10 – How bright do Grow Lights have to be?

A10 – AROUND 100,000 LUMENS.

Light Emitting Diode (LED) Lamps:

With **LED** Lamps, we are looking at a lamp that uses a material that emits light when a current is passed through it.

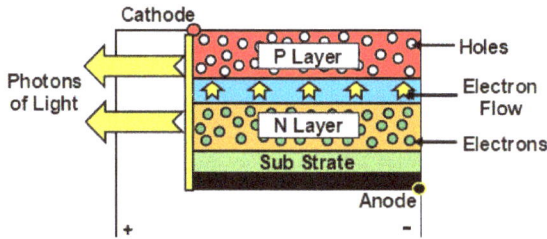

If we go back to the theory of PN Junctions, when we talk about PN junctions we are looking at solid state electronics. Let us make our material as being Silicon or Aluminium: we divide the material into two parts and 'dope' the parts to make a 'P' (Positive) side containing 'holes' and an 'N' (Negative) side containing excess electrons.

The electrons can pass across the joining section to fill the holes, but cannot go back so this makes our PN junction a diode, which will only pass current one way. If we did not put a voltage onto the junction at this point then the electrons would move across and fill the holes and the material would be balanced with no current flow and no charge. Now that we have explained the **'diode'** where do we get the light emitting section from?

What happens when a voltage is applied to the PN junction diode? The PN junction has a phenomenon in that it emits light when a voltage is applied and when electrons flow in one direction (being a diode). This is sometimes called **'electroluminescence'**. In effect the filament solid emits the light without reliance on a gas.

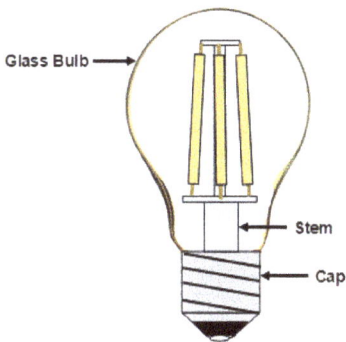

One of the best features of LED lamps is that, like the Incandescent Lamp, there is no warm up time: the full brightness is there immediately. Unfortunately they are more expensive than the equivalent compact Fluorescent Lamps. There is also no choke, instead a small solid state circuit is required.

There has even been a liquid cooled LED lamp developed where the actual bulb is full of a heat conducting liquid, which will in actual fact divert the heat away from the actual LEDs. The design keeps the liquid circulating in the bulb. This is a recent new technology though so maybe not yet in common use.

Similar to Compact Fluorescent Lights (CFL), LED lighting does not appear to have a power factor correction. If that is the case then more current would be needed for the same amount of watts.

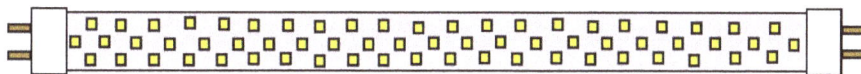

Fluorescent tubes can easily be replaced by **'Ballast Compatible'** LED tubes in the same fitting. These are Type 'A'. With Type 'B' Tubes they do not require a ballast, so a modification will be required to remove it. If you leave the ballast in with Type 'B' it will just use more electricity. LEDs also emit a different light to other lamps, called a 'blue' light. With most LED Lamps you cannot use a normal leading edge light dimmer; it would have to be a trailing edge dimmer or Pulse Width Modulation. Also you need to ensure that the LED in question is actually dimmable.

Manufacturers of Atex Certified i.e. Exd (Flameproof) or Exe (Increased Safety), fittings should be consulted before lamps are changed to LED. For example, fluorescent tubes being replaced by LED tubes. There may be a difference in temperature which would affect the Temperature Classification T1-T6 and with Type 'B' tubes you will be removing the Ballast (Modification).

LED Lamps are the becoming a very popular lamp, but they are maybe not as innocent as they seem. One question I am always asked is, when LED lamps are faulty can I throw them in a dustbin? The answer is definitely **NO!** LED lamps may contain: Arsenic, Barium Oxide, Bauxite (Alumina), **Boron, Copper, Indium, Lead, Manganese, Nickel, Phosphate, Phosphor, Silica, Selenium, Soda Ash, Tin and Zinc**. These materials should not appear in dustbins or skips.

Sum up Questions on LED Lamps:

Q1 – What does LED stand for?

A1 – LIGHT EMITTING DIODE.

Q2 – Can I directly change fluorescent tubes for LED tubes?

A2 – YES: But remember, Type 'B' LED tubes will not require a choke/ballast, as they do not need current control. If the fluorescent is certified Exd – Flameproof, Exe – Increased Safety or Exn – Reduced Risk, you will require advice from Manufacturers and possibly Notified Body Documentation. Type 'A' Tubes will still require the ballast so can be changed direct.

Q3 – I have a circuit that is controlled by a standard leading edge dimmer, can I directly change the Incandescent Lamps for LED lamps?

A3 – NO: It is possible to obtain dimmable LED lamps, but these would probably be specialist and not the type usually available from a supermarket shelf. You may have to change the dimmer for trailing edge dimmer.

Q4 – When I am carrying out Lux/Lumens brightness testing with a Lux Meter does it matter if the lamps have been changed to LED lamps?

A4 – YES: Normal Incandescent Lamps & Fluorescent Lights give off what is called **'White Light'** (nothing to do with warm light etc.) and this is what your standard Lux Meter is designed to detect. LED lamps give off what is called **'Blue Light'** which would register as an inaccurate reading on a standard Lux Meter.

Q5 – Are LED Lamps more efficient than Incandescent Lamps?

A5 – MOST DEFINITELY: Incandescent Lamps give off **90% - 95%** of their energy in **heat** and only around **10% - 5%** in light which is why they are being phased out.

Q6 – Do LED Lamps last longer than Incandescent Lamps?

A6 – YES, DEFINITELY: Incandescent Lamps last around **2000** hours, LED lamps last around **50,000+** hours.

Q7 – Will LED Lamps cost more to buy than Incandescent Lamps?

A7 – YES: Initially this may be the case, but over time you will get more value for your money.

Q8 – Would Incandescent Lamps cost me more in energy over the year than LED Lamps?

A8 – MOST DEFINITELY: The LED Lamp would only need to be under ¼ of the wattage of an Incandescent Lamp to give the same brightness.

Q9 – Can I throw LED lamps in dustbins?

A9 – ABSOLUTELY NOT! Many lamps contain Phosphor which is toxic. In some older LED lamps **Lead, Nickel, Arsenic and more nasties** are used to generate effects like colours. We do not want these in the water or food chain. Also if you break an LED Lamp it is not just the broken glass you have to worry about.

Q10 – Can I obtain LED torch bulbs?

A10 – YES: Torch bulbs also come in LED style. If being installed in a Certified Safety Torch instead of the Incandescent bulb, the Manufacturers should be consulted.

High Voltage Neon Signs:

Back in the past, certain shops such as chemists would have a sign in the window displaying their name. These signs were high voltage with a few thousand volts at their disposal. Shop assistants would be in danger when cleaning the window without isolating the sign.

The signs would usually be filled with a Noble Gas meaning it will undergo a chemical reaction when a voltage is put onto it. These days of course, we can achieve the same as the above without the high voltage using LED (Light Emitting Diode) lighting and a phenomenon called **'electroluminescence'** although this may be more of a harsh light.

The signs were Neon filled at low pressure: Neon being a Noble inert gas with an electrode at either end. By putting a high voltage onto the sign it would make the Neon gas glow and the predominant colour would be red. The only way to bend the electrons and hence light, around the sign in the latter days was, as mentioned, to use high voltage which could be as high as 15,000 Volts!

The electrons flow from one electrode to the other in the tube causing the ions and electrons of the gas in the tube to collide with each other, enhancing the reaction and giving off photons which is where the light comes from. You would sometimes see parts of the sign that would not be lit and this would be caused by that section burning out.

Neon was the most popular gas, hence many songs about Neon lights! Neon light is reddish orange, but other gases could be used or mixed with the Neon gas to give a different coloured light. A Phosphor coating inside of the tube could also alter the colour.

Below are the gases used instead of, or to be mixed with, the Neon gas and the colour they give off when a voltage is applied. Remember, most of them will be Noble gases except for Hydrogen and Mercury. The only Noble gases not to be used are **Radon (Rn) and Oganesson (Og)** as these two gases are radioactive.

Argon (Ar): Noble Gas. Colour: Lavender Colour. This is a very common gas for lighting. It is in the old incandescent lamps as well as fluorescent tubes.

Helium (He): Noble Gas. Colour: Orange. Good thermal conductor. Also used in LED filament bulbs to conduct away a lot of the produced heat.

Hydrogen (H): Colour: Red. Even though Hydrogen is not a Noble inert gas it has been used in Incandescent & Discharge lamps. Used under another name of Deuterium it was used in Ultraviolet lighting where the hot filament causes the Deuterium to emit the Ultraviolet,

Krypton (Kr): Noble Gas. Colour: Green. Similar to Xenon this gas is used in photographic lamps. Also you will find that this gas is in other lamps mixed with other gases.

Mercury (Hg) (Hydrargyrum): Colour: Blue. Another name for Mercury is Quicksilver. It is conductive and very toxic. Mercury easily vaporises when arcs are struck in lamps. Mercury contributes to the efficiency of the lamp, in say, a fluorescent tube and increases the life expectancy. Compact fluorescent lamps used extensively in homes these days contain small amounts of mercury as do many more lamp types. Mercury blended lamps need Mercury for its' design. Not environmentally friendly.

Neon (Ne): Noble Gas. Colour: Reddish-Orange. As above is used in compact tubes when used for signage.

Xenon (Xe): Noble Gas. Colour: Blue. Used in photographic flash lamps. Movie projector lamps still use Xenon arc lamps. Another use for this gas was in HID Halogen auto headlights.

Sum up Questions on High Voltage Neon Signs:

Q1 – Are Neon signs dangerous?

A1 – **NO:** Not if installed correctly, although they are high voltage, they should be safe.

Q2 – What voltage is classed as 'high' voltage.

A2 – **OVER 1000 VOLTS:** The signs that used to be in shop windows could be as high as 10 – 15,000 Volts to coax the outer electrons from their atom. These days with LED lamps we can achieve the same effect without the high voltage.

Q3 – Why is there such a high voltage required?

A3 – **TO IONISE THE NEON GAS:** This was the only way, at the time, of bending light around a series of tube shapes. Electrons are pulled from their atoms causing ionisation. The atoms become 'excited' as electrons bounce about and hit each other and release Photons.

Q4 – Can there be different coloured Neon Lamps?

A4 – **YES AND NO:** Pure Neon shines Reddy Orange when a voltage is applied. Other colours can be obtained either by changing the gas for another type, or adding a coloured phosphor coating to the inside of the tube.

Q5 – Were Neon Lamps used for display purposes in the 1970s?

A5 – **YES:** But these were very small tubes, sometimes because of their size they were called miniature glow lamps. Just a pilot light or actual number shapes could be obtained.

Q6 – Is the Neon gas under high pressure in the tube.

A6 – **NO:** The gas is at low pressure.

Q7 – Do Neon lights buzz?

A7 – **YES:** Because of the atomic reactions going on inside of the tube.

Q8 – Are Neon lights hot or cold cathode?

A8 – **COLD CATHODE:** If we take anything like a hot cathode, then the cathode is heated first to ionise the surrounding atoms e.g. a cathode ray tube, fluorescent tube. In a Neon tube we are just 'exciting' atoms in a Neon gas so no need for hot cathodes here.

Q9 – Is Neon a Noble gas?

A9 – **YES:** Noble gases are inert gases and sometimes have the name **'aerogens'**

Q10 – Neon is said to be **'monatomic'** what does this mean?

A10 – **CONSISTS OF 1 ATOM:** so if you have an Atomic number 10, it has 10 Protons and 10 Electrons. Neon is a very rare gas on Earth, although common across the Universe.

Q11 – Does Neon gas have an odour?

A11 – **NO:** Colourless and odourless, and can be harmful to humans if inhaled in any quantity, for example through a breakage.

Q12 – Do Neon signs burn out?

A12 – **MOST DEFINITELY:** Sometimes you will see parts of the signs light missing.

Less Common Lamps:

Spectral Lamps:

This is a **High Intensity Discharge Lamp (HID)**. Discharge lamps have an arc tube and rely on gas discharge for light instead of a glowing element.

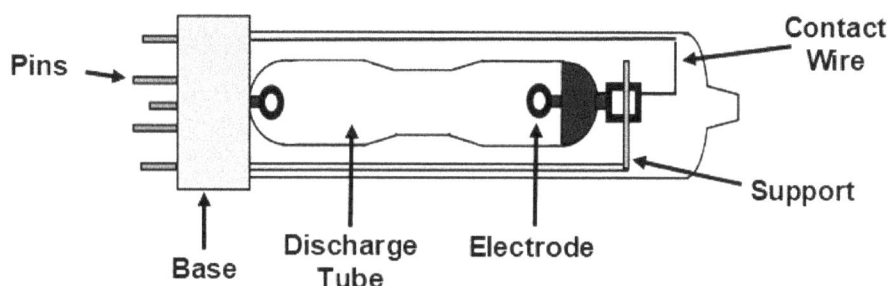

You have probably never heard of Spectral Lamps and they certainly would not be in your hazardous areas they are more likely to be found in a laboratory. They are in fact a Mercury or Thallium Low Pressure Vapour Discharge Lamp, we have talked about discharge lamps earlier in the book. They emit Infrared or Ultraviolet light at a standard level which makes them ideal for calibration.

The above diagram shows an egg timer shaped discharge tube with the two electrodes. Multi-pins are shown on this version, but they can be obtained with an Edison Screw. Other types of Spectral Lamp include Deuterium, Hydrogen, Krypton, Helium, Mercury-Cadmium, Neon and Sodium. Spectral Lamps can be obtained for use in Spectrometry (**Measurement of wavelengths of light**).

Super High Pressure Mercury Lamp:

This is a **High Intensity Discharge Lamp (HID).** (Discharge lamps have a quartz arc tube and rely on gas discharge for light instead of a glowing element.) The temperature can reach around 1200°C.

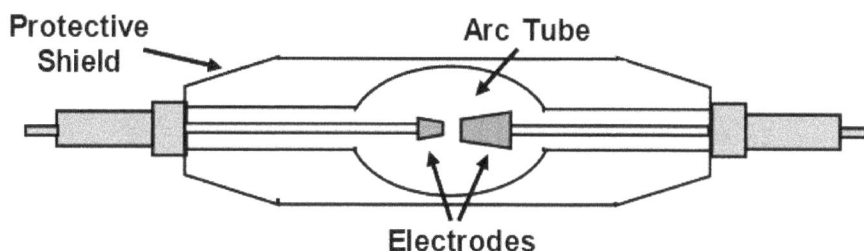

These Discharge Lamps, sometimes called a Short Arc Discharge Lamp, are described as using Xenon, although when the lamp gets to maximum temperature it is actually Mercury gas that is the predominant medium. Super high pressure lamps are used in equipment from car headlights to searchlights. Running voltage is low but arc striking voltage is high.

With the arc tube being under so much pressure there has to be a protective outer shield to stop human injury and removal of this is prohibited if at all possible. **Care must be taken when handling the lamps and protective equipment including thick gloves, eye and body protection is a must.**

Hydrogen Discharge Lamps:

Hydrogen seems to be a fairly volatile gas to use for a discharge lamp but I can assure you they do exist although there is very little information on them.

Sum up Questions on Misc. High Pressure Lamps:

Spectral Lamps:

Q1 – What are Spectral Lamps?

A1 – DISCHARGE LAMPS: Voltage reacts with gas to produce light. Some are not environmental friendly as Mercury Vapour is used, others use Thallium Vapour.

Q2 – Are spectral lamps Infrared or Ultraviolet?

A2 – CAN BE OBTAINED IN EITHER: Specific lamps are available.

Q3 – What is **'Monochromatic'** Lighting?

A3 – LIGHTING WITH A CONSTANT COLOUR AND WAVELENGTH.

Q4 – What are spectral lights used for?

A4 – WHERE MONOCHROMATIC LIGHT IS REQUIRED: Light with a constant colour and wavelength, say for calibration, spectrometers, crime detection etc. Lamps are obtained on what are called **'spectral lines'**.

Q5 – Are Spectral Lamps high pressure?

A5 – YES, BUT ONLY WHILST IN OPERATION:

Super High Pressure Mercury Lamp:

Q1 – What type of lamp is this?

A1 – HIGH INTENSITY DISCHARGE (HID): Sometimes called a 'Short Arc' Discharge Lamp.

Q2 – Being super high pressure are they dangerous?

A2 – NO: There is an outer shield to stop human injury should the arc tube shatter. **If this is removed correct thick PPE and face shield must be worn.**

Q3 – What is the efficacy (Wattage in to light out)?

A3 – AROUND 700 LUMENS/WATT.

Q4 – What type of light do these lamps supply?

A4 – VISIBLE AND ULTRAVIOLET: Phosphors may be applied to make the light visible.

Q5 – Does Mercury make this lamp **NON** environmentally friendly?

A5 – YES: Any lamp that employs Mercury is not environmentally friendly. It is not necessarily the lamp itself, it is the disposal of the lamp and its elements.

Q6 – Where would high pressure Mercury lamps be used?

A6 – PRINTING, PRINTED CIRCUIT BOARDS ETC.

Q7 – What would be a typical wavelength for these lamps?

A7 – AROUND 350nm.

Temporary Lighting in Hazardous Areas:

Lamps:

When looking at temporary lighting in hazardous areas, such as chemical factories, there are certain lamps that, in my opinion, must not be considered, and these are arc lamps such as Mercury Arc/Blended, High Pressure Sodium Arc, and Metal Halide Arc. The reason being for two key reasons:

1 – Arc Lamps have a restrike time. Look at what the temporary lighting is for? Let us consider a team doing work inside a confined space, such as a large industrial boiler. Temporary supplies are run to give them lighting: there is a quick power dip and the lights go out. When the power returns they would be in complete darkness, black, until the restrike time kicks in and then the time for the lamps to warm up to full brightness.

2 – Arc Lamps tend not to like being moved about and knocked. They have built in shock absorber brackets to protect the arc tube against the odd movement, but not robust to withstand the knocks and bangs they would get in temporary supply lighting situations.

LED lighting like Incandescent Lamps would be ideal, they would come back on immediately after a power dip and they would be very bright which is usually what is required.

Certification:

The ideal temporary lighting would be Atex Increased Safety **Exet.** The lamps would be high impact plastic so they would be very hardy and lightweight. Another reason I would choose Increased Safety **Exet** is because the **Exe** protection means that it can go in both Zones 1 & 2 for gases and with the **Ext** protection as well into dust Zones 21 & 22.

The Gas Group of the temporary lighting would be Gas Group 'll' (any gas group) or llC and must cater for dust group lllC. The Temperature Class must be T6 (T85°C).

By taking this approach and not having separate certified lamps for different areas, there is no chance of getting mixed up with the wrong certified lamp in the wrong Zone, which could have consequences.

Cable & Glands:

The cable feeding the temporary lamps would be braided 3 core 1.5mm. The 3 core Live (Brown), Neutral (Blue) and Earth (Green & Yellow) would be required, the braid giving the lamp a lot of mechanical strength which is what is required with temporary supplies. The braid of course is not used as a sole earth which is why the earth core is there.

Glands must be suitable for Zones 1, 21, 2 & 22 just the same as if the temporary supplies were static lamps and must be made as per Manufacturer's instructions. The Universal Gland would be suitable here. As per the certification above, the Gas Group, Dust Group and Temperature Classification should be the highest.

P.A.T. Testing:

Remember these are **temporary leads** so must be given a unique number and PAT Tested at intervals required by regulations, which would be around every 12 months. The test should be formally recorded with readings, date, tester etc.

TN-C & TN-C-S

This type of supply should not be used in a hazardous area because of the combined Earth & Neutral. It could be dangerous on this type of supply to run temporary leads, because if the neutral was to come adrift in the lead it could cause voltage where there should be none.

Sum up Questions on Temporary Leads/Lighting:

Q1 – What is Temporary Lighting?

A1 – NON PERMANENT: Sometimes in certain situations such as confined spaces, for example in a large industrial boiler, where there are no fixed lighting, temporary lighting is run out so that work can be done on, say a shutdown.

Q2 – What lamps are **NOT** suitable for temporary supplies?

A2 – ARC LAMPS: Issue with restrike & warm-up time after power dips and too delicate.

Q3 – What lamps **ARE** suitable for temporary supplies?

A3 – LEDs: They come back on immediately and up to brightness in seconds.

Q4 – What Certification should temporary supplies be?

A4 – TOP ZONE, GAS/DUST GROUP AND TEMPERATURE CLASS: Prevents the potential for getting mix ups and ending up using supplies in the wrong areas for their Certification.

Q5 – What 'Protection' is advisable for temporary supplies?

A5 – ATEX Exe/t INCREASED SAFETY: Lightweight & suitable for all Zones.

Q6 – What type of cable is used for temporary supplies?

A6 – BRAIDED ETHYLENE PROPYLENE RUBBER (EPL): Flexibility and mechanical protection.

Q7 – What cable glands are suitable?

A7 – UNIVERSAL: Will fit all.

Q8 – How often do the leads have to be tested?

A8 – EVERY 12 MONTHS MAXIMUM: They should also be inspected each time they have been used.

Q9 – What is a TN-C System and why is it not suitable for hazardous areas?

A9 – NEUTRAL AND EARTH 'COMBINED': Voltage rises in the System Neutral would mean a potential rise in the Consumer Earth, which could be considerably higher than true Earth.

Q10 – Why is it a risk to use temporary supplies plugged/ wired in to a TN-C System?

A10 – LOSS OF NEUTRAL: In the temporary supply it could cause metalwork earthed by the temporary lead to rise in potential.

Q11 – Similar to question 10, why is it dangerous to lose the Neutral?

A11 – METALWORK BECOMES LIVE: The only path back to the distribution transformer star point is through Earth but with the Neutral gone, anyone stood on the ground touching any metalwork will provide the path.

Q12 – Is it always better to use an RCD?

A12 – DEFINITELY: Many factories have their 240V & 415V Sockets fed via an Earth Leakage unit, which although different, achieves the same objective as an RCD. All temporary extension leads should have this safety system as back up.

Stroboscope/Stroboscopic Effect:

The Stroboscope in a slightly different form, was invented around the middle 1800s by a Belgium scientist called Joseph Plateau. His device was a disc with oblong holes which he looked through and matched to the rotating disc to be measured and so not electronic. This device became known as the **'Phenakistoscope'** The name **'Stroboscope'** came from an invention by Simon Ritter von Stampfer, an Australian inventor who developed his machine at around the same time.

I have known us to measure the speed of a motor with a hand-held tachometer. This has a small wheel that rides on the motor coupling or a pointed probe, which rides in the hole at the centre of a coupling.

So, based on the above, we now have the Electronic Stroboscope to measure the speed of rotation. If we take a rotating disc which could be a flywheel, motor shaft or coupling, which has an unknown speed of rotation, we can find that speed by using this instrument. This may be called the **'flashing frequency'**. The older people around my age will remember that this device was also used to get the ignition timing in car engines as well as used in discotheques.

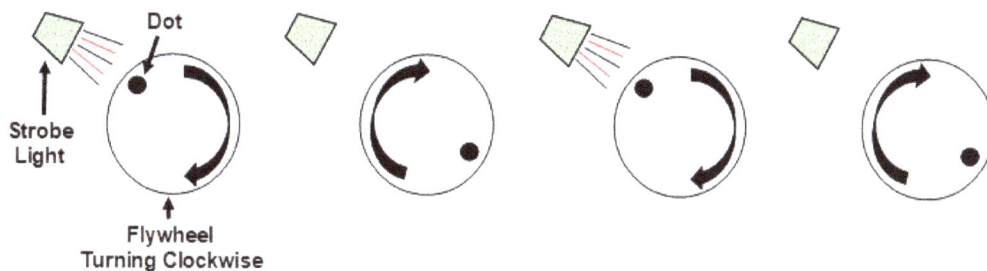

Strobe Light

Dot

Flywheel Turning Clockwise

If we take the diagram above as an example, where we have a flywheel of unknown rotational speed, and put a mark on the flywheel, say, a dot and set the flywheel turning in a clockwise motion. If we adjust the **'oscillator speed'** of the flash of the strobe light to flash every time the dot came round to the same position, the dot would appear stationary. The number of flashes would be the speed of rotation, say 1000 flashes would be 1000 RPM. So, looking at the above diagram, you would see one dot at actual speed, two dots at twice the speed, three dots at three times the speed etc.

Specific types of lamps are used for this because between the flashes the lamp has to cool. So an Incandescent Lamp element would not be suitable as it isn't able to cool fast enough. Xenon, Neon or LED might be more suitable.

It is possible to obtain a stroboscopic tachometer, which will also measure angular velocity that does not require any physical contact with the shaft.

Sometimes in a machine shop, there is a problem with fluorescent lighting giving the same stroboscopic effect and making the rotation machinery appear to be stationary. Using double fluorescent fittings will cut down this effect. Fluorescent lighting frequency can also affect the stroboscope readings.

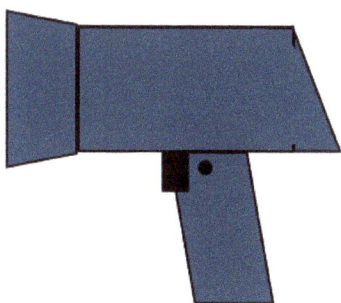

The battery operated Stroboscope instrument today is likely to look similar in design to the instrument in the diagram to the left. The device is operated by a trigger, similar to a gun with 'fine' and 'coarse' adjustments to the **flash frequency,** along with the digital display at the back panel of the instrument. On the front there will be the lamp beads (separate bulbs) giving around 1500 Lux of light. These instruments will usually measure from approximately 50 – 100,000 RPM and are very user friendly. You would need to see the test object for 360 degrees. Prices these days are very reasonable at around £150.

Sum up Questions on Stroboscopes:

Q1 – Is a Stroboscope AC or DC?

A1 – AC.

Q2 – What is a Stroboscope used for?

A2 – **MEASURING SPEED OF A REVOLVING PLANT:** Revolving shafts or discs. They may be used on emergency vehicles to flash the coloured emergency warning lights.

Q3 – Which lamps are the best for this instrument?

A3 – **XENON/NEON/LED:** In the past Neon lamps were used, but these days the lamps will be LEDs. The problem with, say, Incandescent lamps is that it cannot switch on and off fast enough for the frequency of flashes in a Stroboscope. It takes time for the metal filament to heat up and cool down and a fluorescent would take even longer.

Q4 – Are Stroboscopes high voltage?

A4 – **YES, THE STATIC ONES:** This also assists the lamp to flash at a high frequency with the aid of capacitors. Hand held ones will work very similar to a flash in a camera.

Q5 – Can Fluorescent Lighting give a Stroboscopic effect on rotating machinery.

A5 – **VERY MUCH SO:** It is advisable to install double fluorescents in machine shops to cut down this effect. With the mains frequency at 50Hz (Cycles/Second), the fluorescent light actually flashes on and off with the frequency. If we could, with a capacitor, make the second tube flash on when the other is off this would be better.

Q6 – Can Stroboscopes affect your vision?

A6 – **YES:** Any bright light shone into eyes can effect vision. Stroboscopes have another health problem in that they can trigger epilepsy called **'Photosensitive/Photo-induced Epilepsy'**.

Q7 – What actually makes the lamp flash at a certain frequency?

A7 – **ELECTRONICS:** Since the late 1900s.

Q8 – Is it a Stroboscopic effect when wheels seem to stop or go backwards in a film?

A8 – **YES:** Films are a series of pictures fastened together. If the speed of the projector matches the speed of the slides, then wheels will appear to stop or even go backwards.

Q9 – Are hand held Stroboscopes expensive?

A9 – **YES:** Possibly because of the technology required. They can give up to 1500 Lux of light. They can also measure around 50 – 100,000 RPM.

Q10 – Were early Stroboscopes instruments with oblong holes or slits in a disc?

A10 – **NOT QUITE:** The instrument in question was a **Phenakistoscope**. You spun the disc and looked through the holes. When the spin speed of the disc matched the speed of the rotating object it would appear to be stationary. In the early days many were used in amusement arcades and fairs.

Fibre Optic Lighting:

You may see Fibre Optic equipment in your hazardous area in the form of equipment, such as 'Fieldbus', telephones & computer systems, that run the plants. Every day we see Fibre Optic Lighting, it lights up the dashboard switches in our cars, gives us multi-coloured Christmas tree branch ends, domestic fancy lamps and medical instruments such as Endoscopes. Some countries spell 'Fibre' (UK Spelling) as 'Fiber'.

So what are these fibre optics and how do they work?

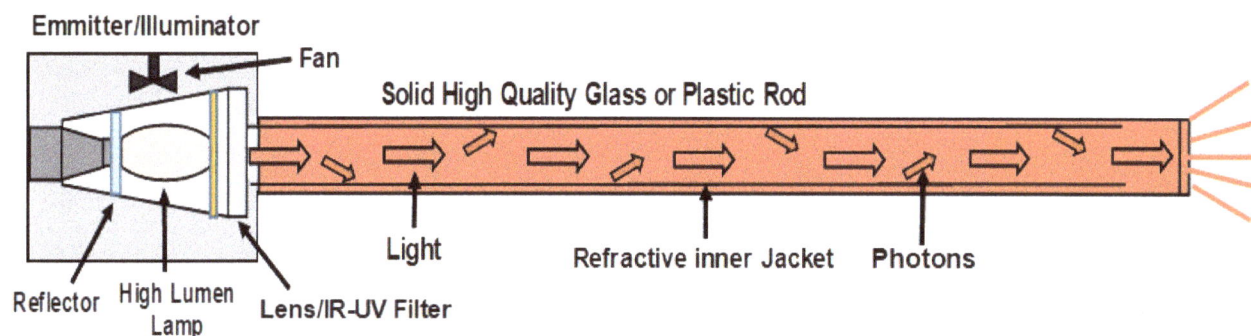

First of all a Fibre Optic Unit is not a lamp on its own, it uses one of the high lumen lamps that we have been talking about, to produce an effect. In the emitter/illuminator we have a very high lumen lamp, similar to a down lamp; behind this is a reflector to transfer all available light forwards; and finally a lens/filter to magnify the light as much as possible. The light is transmitted to a solid, high-quality glass or plastic rod. The plastic rods can be very flexible, as you can see, in the fancy fibre optic lamps.

The diagram above shows one rod leaving the **'light emitter'**, or its correct name is an **'illuminator'**. For ease of explanation the diagram shows one rod, but there can be several rods as in the fancy domestic fibre optic lamps or in fact in your car dashboard switches and instruments. The task of the emitter, as well as magnifying the light, is to filter out any dangerous Infrared or Ultraviolet by putting a filter behind the lens.

Unlike fluorescent tubes which are hollow to hold the gas, this light emitter is solid. Rods that only display light out of the end of the rod are technically called 'End Emitters'. They have a refractive inner jacket that deflects the light inwards, although there is a glow and this is called 'total internal reflection'. Rods can be obtained that emit light sideways, which are technically called 'Edge Emitters' and in this case the refractive inner jacket would be missing.

The rod can range from the width of a human hair upwards. The main bulk of the light comes out of the end. Where I have situated the lens in the diagram, you could have, as many ornamental Christmas trees have, a multi-coloured revolving disc which will alternate the coloured light at the end of the rod(s).

The one point about this lighting system is, that although the lamp unit in the Emitter/Illuminator may get hot, the actual glass/plastic light rods do not as there is only pure light going through them. The actual emitting power unit or Illuminator, depending upon the type and manufacturer, can have a cooling fan built in to cool down the light source lamp.

Telephones Fibre Optics:

With telephones we have a fibre optic cable with several hundred fibre optic strands. These strands are sometimes much thinner than a human hair and can carry many thousands of signals in each strand so imagine how many calls can be handled in the whole cable.

As the diagram at the top shows, some of the signals pass down the solid core and others bounce off the sides because of a refractive inner jacket. The way the light travels down the solid core is called a 'Mode'. A decoder of some kind, such as an optical transducer, needs to be employed to turn the light signals back into electrical impulses.

Sum up Questions on Fibre Optic Lamps:

Q1 – Is the Fibre Optic Lamp a lamp in its own right?

A1 – NO: The lamp can be any high lumen output lamp. Light is transferred to the fibre optic rods or strands (optic fibres) from the lamp housing called the Emitter or Illuminator. These rods or strands are solid, not hollow like a fluorescent tube, and are called single 'modes' or light paths. The benefit with these is that we can bundle a lot of these strands together and make a cable (multimode).

Q2 – Does light come out of the side of the rod or strand?

A2 – YES AND NO: Where the photons bounce off the wall of the rod/strand and light comes out of the end (total internal reflection) of the rod/strand it is called an **'end emitter'** which is the most common. Where light comes out of the side if the rod/strand it is called an **'edge' emitter'**.

What makes the 'end emitter' different is the whole rod or strand is surrounded by what is called **'cladding'** and this is protected by a **'buffer tube'**, and finally an outer jacket, which in industry, may be a range of colours for different modes (consult Manufacturers).

Q3 – What are the uses for Fibre Optics?

A3 – MOTOR VEHICLES, MEDICAL CAMERAS ETC.: The door switches of a car are a good example for motor vehicles and cameras used for keyhole surgical procedures. There would be no internet if it were not for Fibre Optics.

Q4 – How thick are the rods/strands (optic fibres)?

A4 – AS THICK AS A HUMAN HAIR UPWARDS: It depends on the requirements for the Fibre Optic.

Q5 – Do they use Fibre Optics for telephones & chemical plant signals?

A5 – YES: A strand as thick as a human hair can carry 20,000+ signals which makes this system very suited in industry settings where the control room may be some way from the plant. If I wish the signal just to go one way, then that would be called a **'Simplex'** signal and if I wanted the signal to go and return then that would be a **'Duplex'** signal.

Q6 – Can electronic signals be sent by light?

A6 – YES: You would require an optical transducer unit to decode the light signal to electronic signals.

Q7 – If I wish to join one Fibre Optic Cable to another is this possible?

A7 – YES: But alignment is critical and is done under a magnifying unit involving an operation called **'fusion splicing'**.

Q8 – Can I obtain mechanical connectors for Fibre Optic cables?

A8 – YES: Locking Barrel Connectors are available.

Q9 – Can I make the bends in my fibre optic cable too acute?

A9 – MOST CERTAINLY: In this case there will be loss of light signal. Manufacturers will give parameters of angles as cables go around bends.

Car & Torch Bulbs:

Bulbs or lamps? I think we can safely call these bulbs, although some Electrical Technicians may argue otherwise. These are a much smaller version of the Incandescent Lamp we talked about earlier in the book. They are very voltage sensitive! Car sidelight and indicator bulbs for instance will be 12 Volts, but a torch may only be 3 Volts. Any less voltage and their brightness will be affected; any more will cause them to be brighter, but decrease their lifespan.

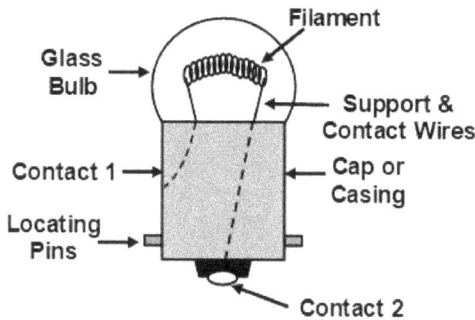

The bulb diagram to the left is what is called a Single Filament bulb and is the type you commonly find in a car or standard torch. It is bayonet cap, although they can be obtained with a screw thread. One contact is the metal body or cap of the bulb and the other contact is at the bottom of the cap. These lamps are universal fitting, which means that it does not matter which way up it is fitted. As mentioned above the lamp will be voltage sensitive and it will display the voltage on the side along with the wattage. They are fairly sturdy, but the filament will become more fragile with age.

The above bayonet cap bulb does not care which way it is inserted into the lamp holder as there is only one filament so the locating pins on the cap are opposite each other. Screw threaded lamps obviously can only be screwed in one way so they will be Single Filament Bulbs. Even these bulbs will eventually become LED, not necessarily because of heat, but more because of a longer lifespan.

The diagram right shows another type of bulb. This one has a Double Filament, the type that you might put into your rear car lamp fitting and serves as sidelight and brake light all-in-one bulb. Again, one contact for both filaments is the metal body or cap of the bulb and at the bottom you can see two contacts, one for each filament.

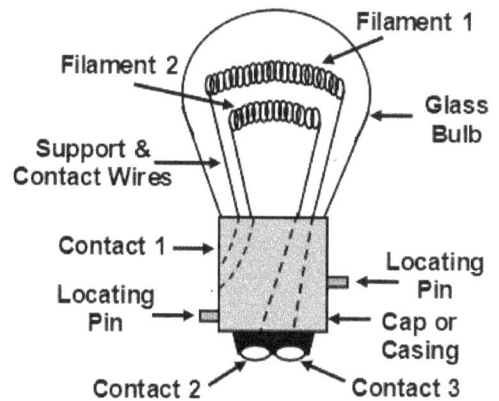

In this instance, although the bulb can be fitted at any angle it does matters which bottom connection is for the brake light and which is for the sidelight, as they will be different wattages/brightness. The locating pins in this case are staggered so that you can only fit the bulb one way in the holder.

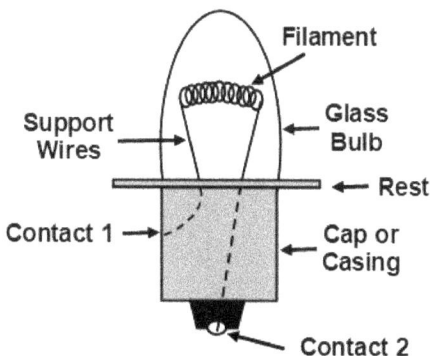

This type of bulb (shown left) has not got a bayonet cap fixing or a screw fixing. The bulb is put into another holder up to, what I have called, the 'rest' in the diagram, and that holder is then screwed into the reflector. On older safety torches this ensured that the lamp was completely encased in the 'O' ring seal, reflector and glass section, so if gas did penetrate the torch it could not get into this section. Again these bulbs are voltage sensitive, details of which will be stamped on the side. Of course in a safety torch only the bulb recommended by the manufacturers can be used AND NO OTHER.

Because we have a miniature Incandescent Lamp here, the element will be Tungsten and the inert gases inside of the bulbs can be Argon, Krypton or Xenon (as in the Incandescent section). Many car lamps used Halogens i.e. Iodine or Bromine (these Halogens are mentioned in the Halogen Lamp section) even Metal Halide has been used in some more expensive bulbs.

The brightest flashlights are LED and go up to 100,000 Lumens, and they may set you back a few hundred pounds. With that amount of light, as you can imagine, there will be heat so the flashlight has to have a built in heat dissipation system. It is unknown whether or not these can be obtained with an Atex Certification.

Sum up Questions on Car & Torch Bulbs:

Q1 – In a car or torch are they called lamps or bulbs?

A1 – **BULBS:** I think we can safely call these bulbs.

Q2 – Can bulbs be dual filament?

A2 – **MOST CERTAINLY:** An example would be the side light/brake light on a car. In this case, the locating pins on the side of the cap would be offset so that the lamp can only be inserted one way to get the right filament to do the right operation.

Q3 – Are torch/car bulbs like the lamps used for house lighting?

A3 – **YES:** They work very similar to an Incandescent Lamp where there is a metal Tungsten filament with a bulb full of inert gas.

Q4 – What is the gas inside of the bulb?

A4 – **USUALLY ARGON, KRYPTON OR XENON:** Some car lamps used Halogens i.e. Bromine or Iodine. Modern lamps will be LED.

Q5 – Can I fit any bulb into an Atex appliance?

A5 – **CERTAINLY NOT:** The manufacturers design of lamp is the only one that can be fitted.

Q6 – What does Atex stand for?

A6 – **Atmosphère Explosibles:** The equipment should also have a Hexagon with Ex in the middle.

Q7 – What 'Protection' are safety torches?

A7 – **INTRINSICALLY SAFE - Exi:** The torch would be encased in plastic including the lens with no exposed metal parts. They must have a safety system whereby the circuit is broken before the torch is opened thus exposing the batteries and internal circuits.

Q8 – Can I change the batteries in my safety torch in a hazardous area?

A8 – **DEFINITELY NOT:** You should not even be carrying spare batteries in a hazardous area.

Q9 – What Zones can Safety Torches be used in?

A9 – **ZONE 0, 1 & 2:** In many cases it will be displayed on the torch. Many can only be used in Zone 1 & 2 and some in Zone 2 only.

Q10 – How bright can Safety Torches go up to?

A10 – **UP TO 300 LUMENS:** Depends upon the type of torch: smaller torches may be under 100 Lumens.

Q11 – Are the batteries special in Safety Torches?

A11 – **NO:** The batteries should be as per Manufacturer's recommendations, which in the smaller torches will be AAA. Ensure that you do not mix batteries in safety torches as it is possible to get an explosion inside the torch if you do. We have had a couple of incidents!

Light Dimmer Switches:

We have discussed several different types of light switches from 1, 2, 3, 4 & Multi Gang with one Way, 2-Way or intermediate switching combinations, single and double pole. Let us now look at how we can dim office and control room lighting. Our light switches, up to now, do one of two things - either switch the light fully on or switch the light off. Some control rooms these days use much softer lighting conditions. You have to be careful as there are lights that cannot easily, if at all, be dimmed. i.e. some LED lamps, CFL and fluorescent tubes have to have special dimmers in some cases although sometimes they are built in. Dimmers can be **Leading Edge** (Triac - see below) or **Trailing Edge** (adjusts the current at points on the Sine Wave before going through '0'). There are such things as **'Inductive'** dimmers which go on the secondary side of a transformer.

Older Dimmers:

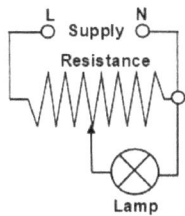

Old light dimmers would work with a varying resistor (a bit like a small autotransformer). So the resistors' wasted heat output was included in the efficiency. This type of dimmer is fine for metal filament lights such incandescent lights and Halogen lights, but they have a problem with Fluorescent, LEDs & HID lamps which may not like voltage drops.

Modern Dimmers Thyristor & Triac:

These days there are several electronic devices we can use to dim lighting called Thyristors & Triacs which put into basic language chop the voltage sine wave at varying points from the base up to its peak and so limiting the power. Without going into too much detail they work on a PN junction technology. A Triac is just two Thyristors in reversed parallel as the diagrams right.

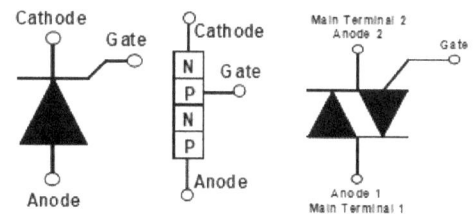

Silicon Controlled Switch (SCS):

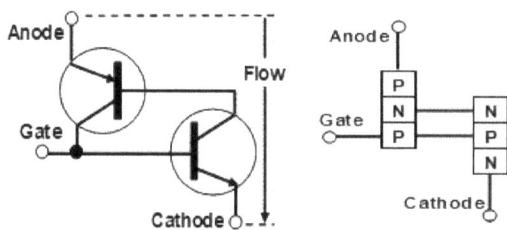

On the diagram (left), I have drawn the Silicon Controlled Switch (SCS) which is based on transistors. Put simply they are electronic switches. The beauty of these devices for controlling the light intensity is that there is no heat and wasted energy. Again these dimmers may not be suitable for several types of lamps.

Pulse Width Modulation (PWM):

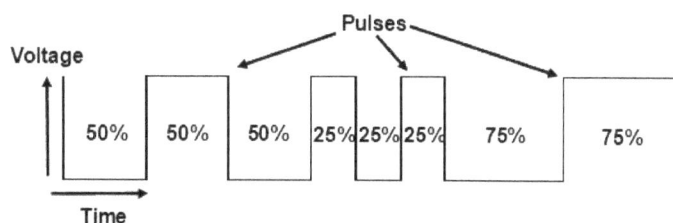

Pulse Width Modulation (PWM) we can say affects the frequency of the supply and, as well as lighting, can used to control the speed of electric motors by enabling them to use their power without producing heat. As the name suggests these devices control the current by a series of pulses. If used for lighting, the longer the pulse the brighter the light. It sometimes is used on LED lighting and other types such as fluorescent, but Manufacturers advice will be required.

Advice can be obtained from either the Manufacturers of the lamps or from the Manufacturers of the Pulse Width Modulation Units. It would be a very expensive exercise to try a 'hit-and-miss' approach to see if it works.

Sum up Questions on Dimmers:

Many people think that if they want to dim the lighting, they just change the standard switch to a standard **'Leading Edge'** dimmer; or if their Incandescent Lamps go and they are already controlled by a Leading Edge dimmer, they simply change them for LED lamps, but this is not always the case.

Remember, in certain lamps you are not only affecting the supply to the lamp, but also to the choke/ballast/electronics, which in actual fact, is trying to keep down the current. Many lamps are sold 'dimmable' so that is not the problem, but that would be displayed on the packet.

If used incorrectly standard Leading Edge dimmers can cause a range of phenomena to do with lighting levels - from going out prematurely as the dimmer is operated, flickering badly, not responding to the dimmer at all, getting brighter as the dimmer is moved down or having a very bad stroboscopic effect.

Q1 – Can I use a standard Leading Edge dimmer on a standard LED Lamp?

A1 – NO: Unless the dimmer is compatible with LED, but if this is so, then it would state on the packet or documentation. Trailing Edge dimmers work best with compatible LED lamps.

Q2 – Can I use a Leading Edge dimmer on Fluorescent Lamps of Compact Fluorescent Lamps (CFL)?

A2 – NO: Standard Fluorescent lighting does not like you to use a standard Leading or Trailing Edge dimmer. I say **NO,** but it is possible to dim fluorescent lighting, but doing so is a complex operation.

You would end up with not enough energy for the tube arc, and the fitting would go into a mode where it is constantly trying to start.

You may require a special dimming regulator along with special cable to the fitting. Ensure the fitting is dimmable in the first place as older fittings may not be. Contact Manufacturers for advice.

Q3 – Can I use a standard Leading Edge dimmer on HID Lamps?

A3 – IT IS POSSIBLE BUT NOT COMMON: If we look at this with the electrical knowledge that we have, discharge lighting has no metal filaments like an Incandescent lamp or Halogen lamp, so varying the voltage or frequency here could be catastrophic or at least produce inefficient light.

We have a ballast and its job it is to keep the current in the arc tube down, if we drop the voltage, that current in theory could in fact go up and destroy the arc tube. Again we could investigate altering the waveform using Pulse Width Modulation. Manufacturers will give advice.

Q4 – Can standard Leading Edge dimmers be obtained multi-gang?

A4 – YES: We had dimmers in one smaller control room which were 4-gang pushbutton. The twist knob dimmer can also be obtained multi-gang.

Q5 – Do I use less electricity with a standard dimmer?

A5 – THE ANSWER HERE WOULD BE YES: You actually are reducing the energy to the lamp as well as the lamp lasting longer.

Common Lamp/Light Bulb Shapes:

Do they call them lamps or light bulbs? As a young apprentice, I was always reminded that bulbs grow in pots and the correct term was lamps for the full unit. However the bulb outer glass shape is called the bulb. Well our 'lamps' as we have seen previously in this book come in many different types i.e. Incandescent, Sodium, Mercury arc etc. They also come in many different shapes and sizes. You have to be careful when renewing lamps that you follow the installation instructions as to vertical cap up, vertical cap down, horizontal or universal. Get this wrong and you will dramatically cut down the lifespan of the lamp. Let us look at a few common shapes and what type of lamp would be in this form?

Traditional Incandescent Lamps:

As we have talked about in the book, these lamps are the traditional Incandescent lamp that has now disappeared from the shelves of our supermarkets. Very inefficient as 90% of the lamp energy emerges as heat and only 10% as light. Still used in health lamps and reptile tanks where heat is required. Bayonet cap or Edison screw.

Elliptical HID Lamps:

This shape can be a number of types. The most common is the High Pressure Sodium Lamps, for instance the Son-E stood for 'Elliptical' whereas Son-T were 'Tubular'. Mercury Arc/Blended lamps also come in this shape including Metal Halide. The actual bulb part of the lamp can have a Phosphor coating on the inside which tends to make the light a bit softer and not so harsh or it can be clear. Also called **High Intensity Discharge (HID)**

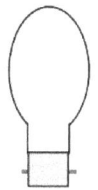

Tubular Lamps:

Son-T is a typical example, the 'T' being for 'Tubular'. Sodium Lamps such as Low Pressure Sodium (SOX) Lamps. Internal or external ignition very important as is installing vertical cap up, down etc.

Compact Fluorescent Lamps (CFL):

Several different shaped tube types with these CFL Lamps. They have a very long life span and low energy lighting. Much more efficient than Incandescent lamps, as not so much energy turned into heat. These contain several grams of Mercury as well as the white toxic Phosphor inside coating, so they should not be disposed of in dustbins. Works on a similar principle as a fluorescent tube.

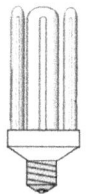

Fluorescent Tubes:

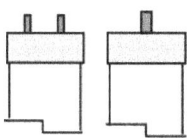

On the left I have shown the ends of two types of fluorescent tubes. The furthest left is what is called a bi-pin tube and is the most common both in Atex and domestic fluorescent fittings. The nearest one being a mono-pin tube which used to be in Exe Increased Safety fittings of the past and is called **'Cold Cathode'**.

Globe HID Lamps:

Globe shape lamps are quite common in domestic lighting, from Incandescent & Halogens to LEDs. However in industry Metal Halide Lamps (HID) tend to be 'globe' shaped. This light may be a bit harsh for domestic use. Much larger than golf ball lamps.

Downlights (These days LED):

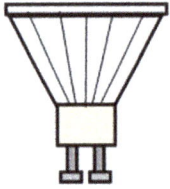

These are used in the domestic arena in bathrooms, kitchens etc., but they are starting to be used quite a lot in industrial offices. They fit subtly into ceilings and look quite effective. A special sucker type tool is required on the type in the diagram on the left, for removal and refitting, but some just have two straight pins with a circlip holding them in position. Can be Halogen or LED. Early problems with heat above ceiling.

LED Tubes:

Light Emitting Diode (LED) lighting is the modern trend. LED lights come in all shapes and sizes and are very efficient, as the majority of their energy comes out in the form of light and not heat. The LED tubes fit into fluorescent fittings with a slight modification as **SOME** of the LED tubes do not require a ballast (choke). They come in white-light for corridors and stairways and warm-light for offices.

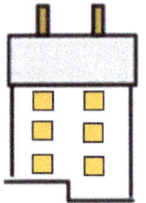

LED Traditional Lamps:

As above, Light Emitting Diode (LED) lighting is the modern trend with the lamps coming in all shapes and sizes most of which are an exact replacement. Most of them are very efficient, as most of their energy come out in light and not heat.

The LED Standard Lamp is designed to be similar in shape to an incandescent lamp and fit into the same situations. These lamps are becoming very popular because of their efficiency.

Spotlights v Floodlights:

Spotlights are used where a more focused light is required on one particular point, but Floodlights tend to spread their light a bit more and may be required where you want to light up a large area. Both of these can be Halogen, Incandescent or LED. Can be obtained as industrial or domestic.

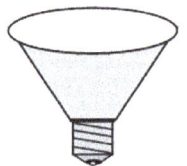

Reflector Lamps:

Some of these lamps have very small reflectors inside the lamp and push a lot of light forward at once. Used in instances such as stage lighting. These are a cross between a spotlight and a floodlight and can be Halogen or LED. Can be obtained in warm light, white light or cool light. Be careful in bathrooms because of the moisture can cause the lamp to explode.

Golf Ball Lamps:

Why they call them Golf Ball Lamps is obvious. They are used in domestic light fittings, festoon lighting and table lamps and certain industrial uses where a lot of light is not required. Can be obtained in a whole range of colours. LED styles of Lamp are taking over as they were based on Incandescent Lamps.

Halogen Capsule Lamps:

Halogen Capsule Lamps are still used in equipment such as projectors and in the domestic Line Lights over bathroom mirrors, where bright, easy to replace, lighting is still required. Capsule gets so hot that it has to be made from quartz, because if pure glass was used the heat would destroy the glass and would be affected by body oils, so no skin contact.

Edison Screw Lamp Caps:

There are many sizes of Edison Screw Lamp Caps ranging from American sizes to European. Some look very similar but they **will not** interchange. Below are diagrams showing the types and an explanation as to where they may be used. The 'E' numbers refer to the diameter of their Edison Screw. Remember that American Lamps can be 120Volts and not 240Volts.

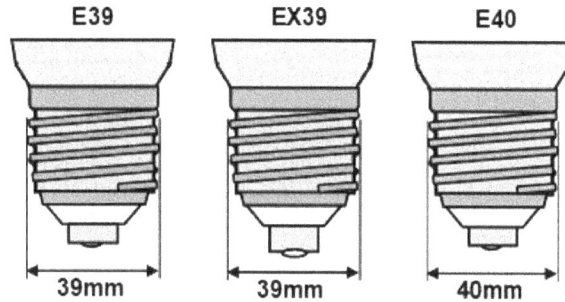

E39	EX39	E40
39mm	39mm	40mm

Above we have the E40, E39 and EX39 sometimes referred to as **'Giant'** Edison Screw (GES) other names may be **'Mogul'** Edison Screw or **'Goliath'** Edison Screw. As you can see there is a slight difference of 1mm between the E40 so usually they are interchangeable. So why the difference? Well, the E39 series of lamp holder are North American and the E40 series are European. The EX39 has a longer base so the E39 lamp may not work in an EX39 lamp holder, but EX39 should work in an E39 lamp holder (always check with manufacturer). **Remember to check the North American voltage of 120Volts compared with European 220Volts.**

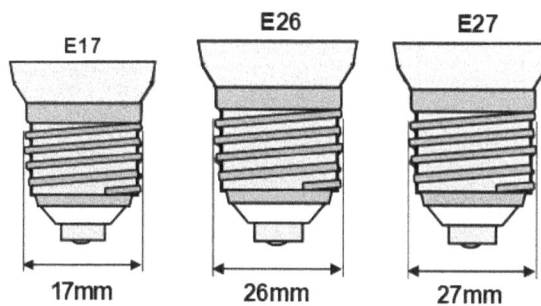

E17	E26	E27
17mm	26mm	27mm

The E17 Edison Screw, 120Volt lamp, above, is not a common size cap and may for instance be the size used in a microwave oven, fridge etc. These are called **'Intermediate'** Edison Screw (IES). When the diameters get smaller the lamps may not be interchangeable, but should be OK with the E26 & E27 with the 1mm difference in diameter. There is another very big difference in the ones above that you need to be aware of: the E26 lamp is **North American 120Volts** and the E27 lamp is the **European 220Volts.** E27 would be the **'Standard'** Edison Screw Cap.

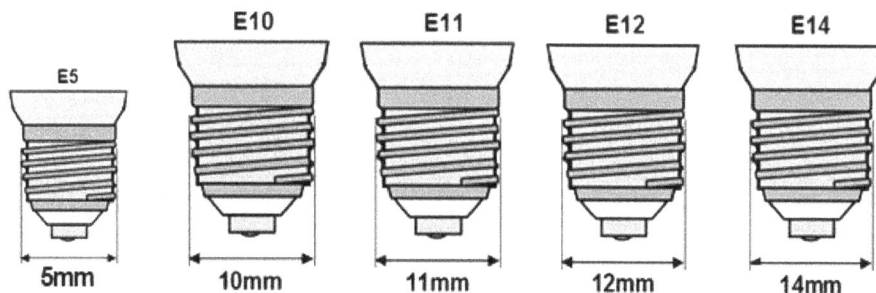

E5	E10	E11	E12	E14
5mm	10mm	11mm	12mm	14mm

The E5 Edison Screw is a very tiny cap and can be called the **'Lilliput'** Edison Screw (LES). These are extra low voltage for models, toys etc. The E10 are called **'Miniature'** Edison Screw (MES) and are again, low voltage and used in torches, bicycle lamps etc. The E11 is called **'Mini Candelabra'** Edison Screw (Mini-CES) and the E12 **'Candelabra'** Edison Screw (CES) and these are North American 120Volt lamps. The E14 Edison Screw is what we term as **'Small'** Edison Screw (SES).

Bayonet Cap Lamps:

Lamps with Bayonet Cap are used extensively in the UK – in homes, industry and cars etc. Below are 4 common sizes of bayonet cap which you may come across.

Bayonet Cap Lamps are usually designated 'BA' with a number afterwards which refers to the diameter of the cap in millimetres. So a BA22 would be a bayonet cap which is 22mm in diameter and is a standard UK fitting.

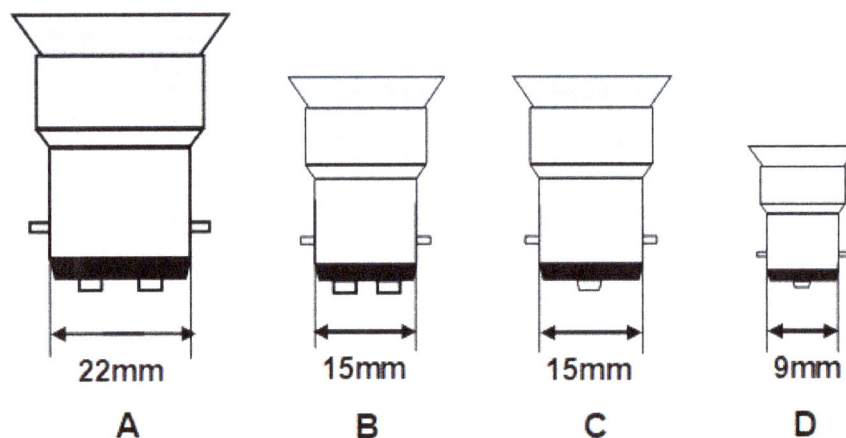

| 22mm | 15mm | 15mm | 9mm |
| A | B | C | D |

Cap 'A' (above) is a very common type of Bayonet Cap, typically used in table and ceiling lamp holders. This cap is 22mm in diameter and has two electrodes on the bottom. It does not matter which way round you insert the lamp into the lamp holder as the location pins on the side of the bayonet fitting are level. his is referred to as a **'Standard'** Bayonet Cap (BC).

Cap 'B' is 15mm and is sometimes used in table lamps and again it does not matter which way round the lamp is inserted as the location pins are level. This fitting also has two electrodes on the bottom. This is referred to as a **'Small'** Bayonet Cap.

Cap 'B' is 15mm in diameter and is used on car lamps with two filaments i.e. a tail light and stop light. So, there are two electrodes at the base, and with this cap the side of the bayonet cap fitting is one connection and the electrodes at the base have one for each filament in the lamp i.e. tail light and stop light. It matters a great deal which way round the bayonet is inserted. The location pins on the side are off-set so that the lamp will only fit one way in the lamp holder. The offset pins also ensure that you cannot fit this cap into a household 240Volt lamp holder. This is referred to as a **'Small'** Bayonet Cap (SBC).

Cap 'C' again this is used in a car light where the lamp only had one filament such as an indicator. The base of the cap in this case only has one electrode, the side of the Bayonet Cap being the other. It does not matter which way round the lamp is fitted in this case, so the location pins on the side of the bayonet are level. This is referred to as a **'Small'** Bayonet Cap (SBC). **Do not get these mixed up with domestic lamps.**

Cap 'D' is the sort of bayonet fitting that you might find in low voltage appliances such at torches etc. and like Cap 'C' there is one electrode at the base and the side of the bayonet being the other. Again it does not matter which way round the lamp is fitted so the location pins are level. This is called a **'Miniature'** Bayonet Cap (MBC).

There are many more bayonet cap types used in Europe and the UK that are not so common. For example, there used to be a type of 25Volt Incandescent Lamp used in older gripper hand lamps, which looked exactly the same as a 240Volt lamp, but the Bayonet Cap had three locator pins out of the side 120 degrees apart and as such it was impossible to fit it into 240Volt lamp holder.

Some Miscellaneous Lamp Fixings:

We have discussed the Bayonet Cap and the Edison Screw Cap which are the most common lamp fittings, but some lamps have different ways of fixing them into the lamp holder. Let us have a look at some examples:

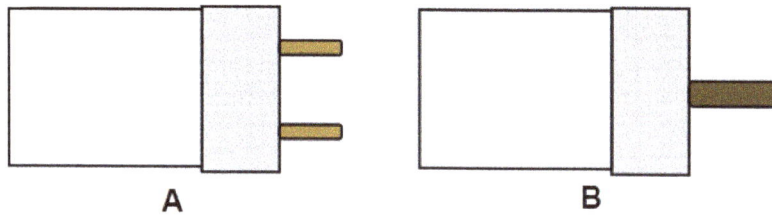

A   **B**

Fluorescent tubes: Above are two types of end caps of fluorescent tubes. The above left 'A' is the common Bi-pin Tubes that are in fluorescent fittings both certified and non-certified these days. There is a heater/tungsten electrode connected to the pins inside the tube, these heaters heat the gas in the tube prior to the strike. (See fluorescent fittings) The above right 'B' is what is called a Mono-pin Tube and used to be in some certified fittings, Exe, of the past. No heaters in this tube, as it just relies on voltage to strike the arc from one tube electrode to the other. Called a cold cathode start. Fluorescent Tubes are being replaced by LED Tubes with same bi-pin fittings, but no ballast.

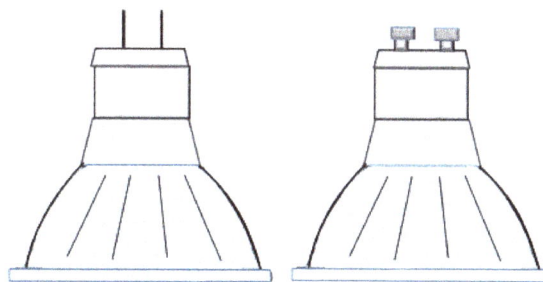

Above are two different fixings for ceiling **'Downlights'**. Above left there is usually a large, easy to remove circlip holding the lamp in place and when removed the lamp can be unplugged from the supply wires. Above right, are a little more difficult to remove and a sucker type remover has to be attached to the lamp to turn it anti-clockwise for removal. It is doubtful that these lamps can be removed with fingers. In the past they would usually have been Halogen but the modern lamps will be LED.

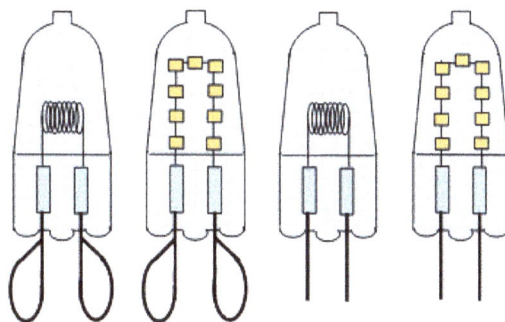

The four lamps above are called **'Capsule'** Lamps and as you can see there are two types of mounting. Both of these mountings were Halogen lamps but the modern ones are LED. These are used in many items in both domestic and industrial equipment. Because of the very high temperature they use Quartz glass. If the capsule was made out of pure glass it would not withstand the heat. With the capsule being made of Quartz and not glass, it is very sensitive about being handled. Your fingers can deposit oils out of your skin onto the Quartz and cause premature capsule rupture when it heats up. Capsules should be thoroughly cleaned if handled. Use the packet as a shield between your fingers and the lamp when replacing.

Common Lamp-holders:

Lamp holders come in all shapes and sizes and in both Edison screw and bayonet cap fixings. They can be plastic or brass depending on the duty and location. Many fittings such as fluorescent and well-glass etc. will have their Manufacturer's lamp holder already fitted and this cannot usually be changed for a different one. Certified fittings such as Exe & Exd certainly **CANNOT** be altered otherwise this will de-certify them.

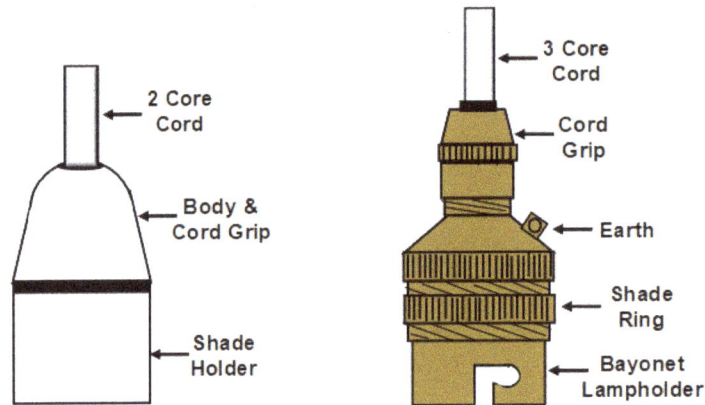

Above, are two common lamp holders. The left one is the familiar plastic lamp holder that is in every home, whilst the one on the right is brass and may be used in instances where the lamp holder has to be sturdier. Both can be obtained in Edison Screw or Bayonet Cap and switched or un-switched. The plastic lamp holders cord grip is designed so that the weight of the lamp and shade is not hanging on the cores of the cord and the brass lamp holder has a clamp which grips the cord. The brass lamp holder will require an earth so the cord must be 3-core. Tthe plastic one has an earth terminal, but is not used in connection with this lamp holder.

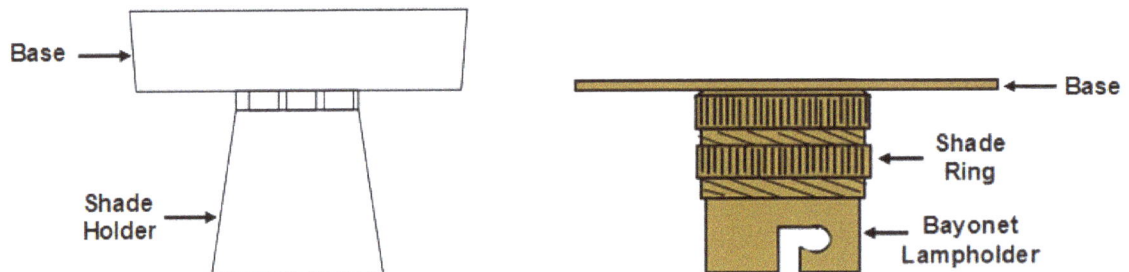

Above are diagrams of Batten type lamp holders and can again be made out of plastic or brass, Edison Screw or Bayonet Cap. It must be remembered that brass fittings must have an earth unless the equipment is double insulated. There will be a facility to run an earth into the base or rose of the plastic holder. The plastic one may be used in a bathroom situation where a cord may invite water to run down the flex and into the lamp holder connections.

Above is an example of **'Festoon'** Lamp Holders. The 'T' shaped lamp holder seals around the cord to a usual level of IP44. Used for outside decorative occasions. Not for hazardous areas.

Lamp Filament Materials:

There are not that many Lamp Filament materials available because of the high temperatures that they have to withstand when working - which could be up to 3,000°C+. These materials also have to be in one continuous length as any joints will not withstand the high temperature so, if we take the Tungsten filament in the Incandescent lamp it would be coiled and the coil coiled and maybe coiled again to obtain the correct length for the filament resistance.

Some materials used for lamp filaments cause blackening of the inside of the lamp. Tungsten in an Incandescent Lamp, especially cap down, actually evaporates when working and collects as a black deposit inside of the glass. This, I believe, is what caused the end of the Carbon filament of the past. Lamp inventors had huge problems with the blackening of the glass, some of which ended their designs there and then.

So what materials are we looking at?

1900s to date:

1. **Tungsten:-** Melting Point around 3,420°C

   Uses: Incandescent, Halogen & Fluorescent lights in the UK.

   Chemical symbol '**W**' as another name for Tungsten is '**Wolfram**'

2. **Carbon:-** Melting Point around 3,500°C

   Uses: More or less obsolete for lighting in modern arc lights in the UK.

   Chemical symbol '**C**' the same as **Graphite.**

3. **Tantalum:-** Melting Point around 2,900°C

   Uses: Rarely in Neon lamps in the UK. (Also many electronic components.)

   Chemical symbol '**Ta**'

4. **Osmium:-** Melting Point around 3,035°C

   Uses: Incandescent lamps in the UK.

   Chemical symbol '**Os**'

   Osram Company name was formed from **Os**mium and Wolf**ram**

1800s/1900s:

As you read through the early pages of history you will see that inventors used a whole range of materials for their filaments, ranging from cheap to very expensive materials as listed below:

1. **Tungsten:-** Used in the early 1900s by William Coolidge. Used in many lamps especially the older Incandescent and Fluorescent.

2. **Carbon:-** was used in the 1800s by Charles Brush & Humphrey Davy.

3. **Copper Coated Carbon:-** was used in the 1800s by Charles Brush in his street lamps

4. **Bamboo:-** was used in the 1800s by Thomas Edison.

5. **Platinum:-** was used in the 1800s by Humphrey Davy. Very Expensive.

6. **Carbonised Paper:-** was used in the 1800s by Joseph Swann.

7. **Carbonised Cotton Thread:-** was used in the 1800s by Joseph Swann.

Gases used in Light Bulbs:

There are many substances, chemicals and gases used in light bulbs, some are inert gases and some are conductive to assist the arc. For example, in the Incandescent lamps the main gas is Argon with a small amount of Nitrogen, both inert and lengthen the life of the Tungsten element. In Mercury Blended lamps, the gas in the arc tube is Mercury. What the light bulbs are made of plays a very important part. The **Seven gases called 'Noble' gases** are, Helium (He); Neon (Ne); Argon (Ar); Krypton (Kr); Xenon (Xe); Radon (Rn) and Oganesson (Og). Noble gases have a very low reactivity and are classed as inert, with the possible exception of Oganesson.

Only Helium, Neon, Argon, Krypton and Xenon are used in lighting. **Radon and Oganesson are radioactive.** We must not forget Nitrogen which although is not a Noble gas, it is an inert gas used in lighting and is 'N' on the Periodic Table.

Let us have a look at some of the substances, chemicals and gases mentioned and their use:

1. **Helium** (He): Inert **'Noble'** Gas. Good thermal conductor. Used in LED filament bulbs to conduct away a lot of the produced heat.

2. **Neon** (Ne): Inert **'Noble'** Gas. Can be used in compact tubes when used for signage. Neon has a red glow.

3. **Argon** (Ar): Inert **'Noble'** Gas. Used in Incandescent lamps, Compact Fluorescent (CFL) and Standard Fluorescent Tubes to increase the life of the filament. Oxygen would corrode the very hot filament very quickly and Tungsten evaporates when the lamp is working, so Argon this would slow this process up. This is probably the most popular inert gas used in lighting.

4. **Krypton** (Kr): Inert **'Noble'** Gas. Similar to Xenon this gas is used in Photographic lamps. Also you will find that this gas is used in other lamps, mixed with other gases.

5. **Xenon** (Xe): Inert **'Noble'** Gas. Used in Photographic Flash lamps. Movie projectors still use Xenon Arc Lamps. Another use for this gas is in HID Halogen Auto Headlights.

6. **Nitrogen** (N): Inert Gas. Mixed with, or instead of Argon, is used in many forms of lighting. Carries heat away from the very hot filament and like Argon above, slows down the corrosion of the filament which would happen if the gas was Oxygen.

7. **Mercury** (Hg) (**Hg because of Latin Hydrargyrum**): Conductive and very toxic. Mercury easily vaporises when arcs are struck in lamps. Mercury contributes to the efficiency of the lamp in say a fluorescent tube and increases the life expectancy. Compact fluorescent lamps used extensively in homes these days contain small amounts of Mercury, as do many more lamp types. Mercury Blended lamps need Mercury for its design. Unfortunately, Mercury is not environmentally friendly.

8. **Halides:** are a compound of a Halogen Atom and another element to form another substance such as a salt.

9. **Halogens:** The full list of Halogens are, Astatine (At), **Bromine (Br)**, Chlorine (Cl), Fluorine (F) and **Iodine (I),** but only Bromine and Iodine are used in lighting. The Halogen lamps can be called Tungsten Halogen or Quartz Halogen. These lamps are Mercury free and the closest to normal sunlight of all lamps.

10. **Sodium** (Na): A soft metal with a low melting point. A good conductor of electricity. Used in Low and High Pressure Sodium Lighting. Has a terrible colour index and items in this light never look their true colours. Again High Pressure Sodium Lamps contain Mercury.

11. **Hydrogen** (H): Even though Hydrogen is not an inert gas it has been used in certain lamps in Spectrometers. Used under another name of **Deuterium** it was used in Ultraviolet lighting where the hot filament causes the Deuterium to emit the Ultraviolet.

Materials used in Lighting & Switches:

1. **Phosphor:** A very toxic powder coating on the inside of glass light bulbs and tubes. Causes luminescence when submitted to energy such as photons. Coats the inside of many forms of lamps the most common being fluorescent tubes.

2. **Glass:** Commonly used to form the bulbs and tubes on most lamps. Lamps that get extremely hot may have to turn to Quartz (see below). Soda-lime glass can also withstand higher temperatures.

3. **Quartz:** Because of the very high temperature of some lamps, capsules made out of pure glass would not withstand the heat. Capsules made of Quartz are very sensitive to being handled because the oils in human skin, which can cause premature rupturing of the capsule. The packet is used as a shield when installing the lamp.

4. **Brass:** Used for many lamp holders and pins. If brass is used, then as a metal, an earth connection is present and an earth core must be provided by the incoming electrical system. Sometimes the lamp cap is made of brass. **An alloy of Copper (Cu) and Zinc (Zn).**

5. **PVC:** Correct name is Polyvinyl Chloride. Used for many lamp holder surrounds as well as many items of equipment in the electrical world. Lamp caps/bases could also be made of brass.

6. **Ceramics:** Used for many lamp holder surrounds. Porcelain is a type of ceramic. **Comes from the Greek language meaning 'pottery'.**

7. **Bakelite:** Very brittle and gets worse with age. Used on many lamp holders and electrical equipment. The correct name is **'Polyoxybenzylmethylenglycolanhydride'. Made from Phenol and Formaldehyde.**

8. **Molybdenum:** A chemical element with the symbol 'Mo'. The very thin wires that support the Tungsten element in Incandescent light bulbs are made of Molybdenum. It remains stable under very extreme heat.

9. **Aluminium:** The lamp cap/base of any Edison screw or bayonet cap lamps are usually made of Aluminium 'Al'. Also used in fluorescent tube end plates.

10. **Copper:** Used in many cases, as the lead wires up to the elements or electrodes. Symbol: **Cu (because of Latin Cuprum)**

11. **Steel/Stainless Steel:** Many modern light switches and sockets have this 'brushed steel' look to make them look modern.

12. **Mercury**: Sometimes called 'Quicksilver', Mercury is used in Fluorescent tubes, compact Fluorescent lamps and Mercury Blended lamps. Symbol: **Hg (because of Latin Hydrargyrum)**

13. **Nickel-Iron**: A material used for the connecting wires inside of a lamp usually dipped in borax to make the wire more adherent to glass. Symbol: **FeNi**

14. **Vitrite:** A type of glass which is very hard, an insulator, coloured black and used around the base of the cap of a lamp to keep the electrodes apart.

15. **Polycarbonate:** A type of transparent plastic mostly used for diffusers. Symbol: **PC (because Polymer with Carbon intertwined)**

16. **Acrylic:** Like Polycarbonate, this material is very strong and transparent. Used for lighting diffusers.

17. **Thallium:** Low melting chemical element. Atomic number 81 on periodic table. Used in Spectral lamps. Symbol: **Tl.**

18. **Lead**: Used in some lighting to generate colours, especially LED lamps.

Information on Light Bulb Packet:

When I order my lights, in the catalogue there are the Compact Fluorescent, Halogen or LED lamp I want, but what exactly am I looking for when I am choosing my light bulb? Well, the information that is on the light bulb packet and may be as follows:

1. **Type of Lamp:** Compact Fluorescent (CFL), Halogen, LED etc.
2. **Type of light:**

    a - **Warm White** is equivalent to the old pearl Incandescent bulbs and good for lights over desk in offices, in domestic circumstances, study areas etc.

    b - **Soft White** will have a slightly whiter look and can be found perhaps over photo-stat machines etc.

    c - **Cool White** is starting to get a bit too harsh for offices, over desks, etc. It's a more bluish light. Maybe good for corridors, stairways etc.

    d - **Daylight** is a very harsh white light as the label says more like daylight. Certainly too harsh light for offices, people can get headaches. More suited to say industrial plant where a lot of lighting is required in dark spots.

3. **Colour Temperature (Kelvin):** From 2700°K – 6500°K. This is quite important as you match the colour of the lamp above to the 'mood' of the location: (These figures may vary slightly depending upon the make)

    a - Warm White   –   Around 2700°K

    b - Soft White   –   Around 3500°K

    c - Cool White   –   Around 5000°K

    d - Daylight   –   Around 6500°K

4. **Brightness in Lumens:** This is a very important measurement as this is how bright the above lighting will be (Sometimes in % as bright as). For instance below is a rough guide:

    a - 100 Watt Incandescent lamp was around 1500 Lumens,

    b - 60 Watt Incandescent lamp was around 800-900 Lumens.

    c - 40 Watt Incandescent lamp was around 500-600 Lumens.

5. **Equivalent Watts (the large number):** This figure, I my opinion, does not give any relevant information and is the equivalent Wattage to the old Incandescent lamps.

6. **Wattage (the small number):** This is the Wattage that the lamp itself takes to give you the light. Again its of little interest.

7. **The Cap:** This is very important, as you need to know if the lamp is going to fit your lamp holder. So what **type** is the cap? Is it Bayonet Cap, Edison Screw or some other? What **size** of Bayonet Cap, Edison Screw or other?

8. **Lamp Life:** On the packet it should give a lamp life time. So it may say something like: if the lamp was left on 3 hours a day, every day, it would last 'X' number of years.

9. **Lamp Contents:** Does it contain Mercury (Unlikely)? Important maybe in the way that you dispose of old lamps. **LEDs contain lots of nasties one of them being Arsenic.**

Measuring Light Brightness/Illuminance:

So, how do we measure light intensity or '**Illuminance**'? Well, we can use a light meter similar to the one in the diagram below:

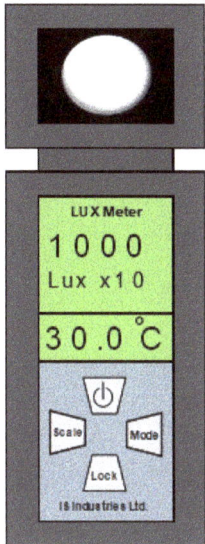

There are many different makes and models on the market. We measure this illuminance in units called '**Foot Candles**' (**Imperial**) or '**Lux**' (**Metric**) which replaced candle power. We need to know how much light is falling on a particular surface, be that a desk in an office, the floor in a corridor or areas of a production plant. Sometimes '**Lux**' and '**Foot Candles**' are mixed up with '**Lumens**'.

You will notice that there is a temperature reading on the Lux Meter to the left. This is important to the photodetector on the instrument itself rather than the reading. The temperature would have to be quite high to affect it.

It is possible to obtain instruments which are capable of storing the information as you go round the different locations and downloading onto a computer. Sometimes over time, lamp units become less bright between readings, fluorescent tubes are a good example, their light can decrease with age.

1.   **Foot Candles** (Imperial) are the number of Lumens falling on a square foot of surface from 1 foot away.

2.   **Lux** (Metric) is the amount of Lumens falling on a square metre of surface from 1 metre away.

This may be called the '**Luminous Flux**' value. So to sum up. Lumens are a measurement of the spectrum of light the human eye can detect. Foot Candles or Lux are the units of measurement of those Lumens falling on a particular surface.

It may be that the '**white light**' instrument above will measure standard office and plant lighting levels, **but a different instrument may have to be used to accurately measure LED lighting** because of the '**blue light**' wavelength. (White light here has nothing to do with white light below)

There are also different types of light namely 'warm light' and 'white light' these are also variables in LED lighting. Warm light for instance should be in an office environment where the light does not want to be harsh. White light can be in corridors or stairways. Putting white light in an office environment would make the lighting too harsh and may cause headaches. Whereas in a corridor or staircase people are not there for very long and these areas can be very brightly lit. If we over light an area we can cause just as many problems as under lighting it.

For the best results, the meter should be stable i.e., placed on a desk, table or surface, if you stand with it the readings may fluctuate. The surface would always give the instrument a stable height for all of the readings.

Very roughly 1 Foot Candle = 10 Lux and therefore 1 Lux = 0.1 Foot Candles. Some **approximate** lighting levels are as follows (Please check standards for current true values):

Open Plan Offices – around 300-500 Lux   **Stairways** – around 100 Lux

Workshops – around 700 Lux   **Corridors** – around 50-100 Lux

Switch Rooms & Substations – around 500 Lux   **Process Plants** – around 100-150 Lux

Control Rooms – 200-500 Lux

Please always refer to National Standards for lighting levels in your companies.

Light Switches (Offices & Control Rooms):

Light switches come in many shapes and sizes from a single switch which only has one circuit to a switch panel with several switches. Below switch terminology is explained I have done domestic first, which may be used in offices, to give an idea as hazardous area switching can be more complex:

Gang:

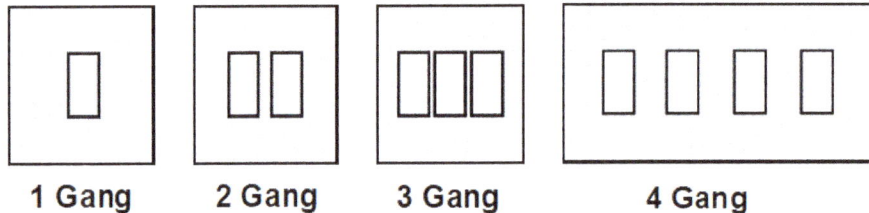

1 Gang   **2 Gang**   **3 Gang**   **4 Gang**

The diagram above shown a 1, 2, 3 & 4 Gang switch where all the switches should be off the same phase & circuit. The 1 & 2 Gang will usually fit into a shallow metal wall box or plastic pattress. The 3 Gang would require a deeper back-box & 4 Gang switches are much wider and would require a deep, double socket, box. The switches in the above diagram can be obtained in 1 or 2 way. Just one point if the switch is 3 Gang and one of them is required to be 2 way then all 3 switches will be 2 way even though two of them just require to be 1 way. Another point in domestic office light switches, is that they are commonly single pole (does not switch the neutral) whereas a hazardous area switch would be double pole (Also switches the neutral).

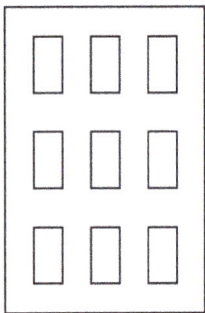

Sometimes if the office or control room is quite large in area and there are a lot of lights to switch, the switch may go considerably higher than the number of switches in the diagram left. I actually came across this example similar to the 9 Gang switch in the diagram. It is possible for the building designers to spread the lighting load over 3 phases. Looking at the diagram on the left the top row, it would be say, the **Brown phase**, the middle row, the **Black phase** and the bottom row the **Grey phase**. This could be dangerous, if precautions are not taken i.e., phases must be segregated inside the switch box and a notice put on the front stating that internal voltage is 415V and not just 240V.

I am not aware if there are any rules or regulations preventing the above example. If not, I am sure that you can see there could be a dangerous situation if the technician was not careful to isolate the lighting. So to work on the 9 Gang switch unit **ALL 3 PHASE FUSES MUST BE REMOVED.**

Way:

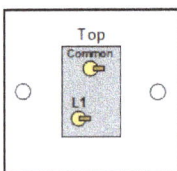

The switch in the diagram on the left is a one or single-way switch so there are only two connections on the back. This is just to switch a light or a circuit of lights on and off. Look for the sign stating 'Top', this is to ensure that the switch is the correct way up i.e., down for 'on' and up for 'off'.

The switch in the diagram on the right is a double or two-way switch so there are only three connections on the back. This switch would be used on, say, a staircase where the lights may be switched 'on' at the bottom and 'off' at the top. Look for the sign stating 'Top' this is to ensure that the switch is the correct way up.

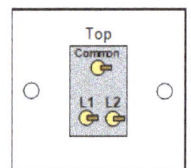

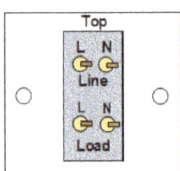

The switch in the diagram on the left is a double pole switch where the neutral is switched as well as the live. There is not much call for this in office lighting and control room lighting, but a must when working on lighting in Zones 1 & 2. Look for the sign stating 'Top' this is to ensure that the switch is the correct way up i.e., down for 'on' and up for 'off'. Make sure not to get mixed up with an intermediate switch with four terminals.

Plant Lighting Panels:

Back in the 1960s to 1970s most electrical equipment was Flameproof. On site all motors would be Flameproof with a Flameproof Starter next to them. ExN Non-Incendive equipment and motor control centres (MCCs) did not come in until later. Lighting panels on the plant would take a very similar form and would also be flameproof, similar to the one below. All switch units were usually double pole, but some were single pole with a small removable neutral link. Of course, removing the fuse and the neutral link would serve as double pole isolation.

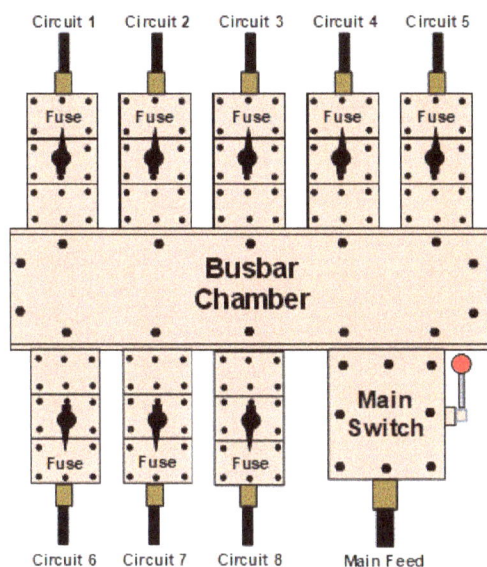

Flameproof lighting panels such as the one in the diagram on the left, were quite common on the chemical plants that I was working on in the 1960s & 70s. The switches would be double pole with the fuse compartment above on the top row and below on the bottom row. When the switch was switched 'on', for safety the lever position stopped the fuse compartment from being opened so that the fuses could not be removed with the lights on. The individual switches could be padlocked in the 'off' position. In the centre of the panel was a bus-bar chamber which fed each switch fuse. The bus-bar chamber was fed via a main switch which would be fed from a **C**ombination **F**used **S**witch in a local switch-room. This main switch could also be locked in the 'off' position. These panels would be completely overhauled/inspected every set number of years. Some plants may still have these, albeit a later version.

Some early lighting panels such as the one above, were physically switched 'on' and 'off' by the plant Operations Technicians, which of course would mean that sometimes lamps were left on unnecessarily wasting a lot of electricity. In most cases, a photocell control was added which means that a contactor would feed the main switch. There are areas under the plant which do require human control and not on photocell because of dark zones that may require light for safety even during daylight hours.

Several drawbacks with these panels were, that the threads would go on the bolts/screws or technicians would sheer bolts during maintenance. Flameproof is not waterproof and older grease was not of the standard of silicon grease, so water could sometimes penetrate the enclosures. Switches would close properly. All of these faults were major faults to repair as the whole panel may have had to be isolated depending where the problem was. Accessories like stoppers in the busbar chamber would be put in from the inside forcing you to isolate first.

Sometimes if it was only a small plant which did not warrant a huge lighting panel, then as many individual Double Pole switches that were required for the circuits would be mounted in a line on the plant and the feed just looped at the bottom. Sometimes Emergency Lighting Circuits would be done this way. As you can see each switch lever is over the fuse compartment so fuses cannot be removed switched on. These were usually Flameproof.

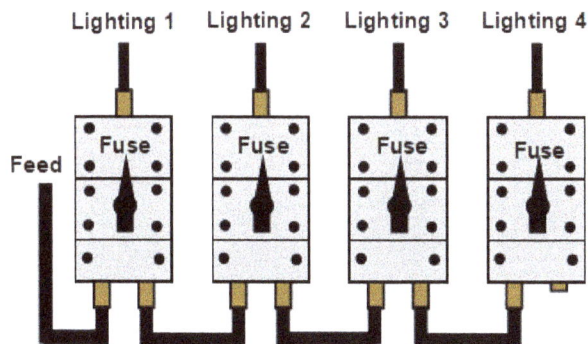

Today, lighting panels will be Atex Exe (Increased Safety) and much more accessible and cheaper than their Exd (Flameproof) counterparts. Instead of having expensive lighting panels out on site in our hazardous zoned area, it may well be very practical and cheaper in the long run to have the lighting panel in an uncertified area such as a switch room and just have small control switches on site. Maintenance would be much cheaper although cabling the lighting circuit would increase initially. The lighting panel in this instance would of course be uncertified.

Light Switches 2- & 3-Way Movement:

2-Way Switching:

If the light is just a single light in a small office room then it is 1 Gang, 1 way, meaning that it switches the light 'on' and 'off' and only has two terminals, but in a stairway or a very large room, say an open plan office, where there may be more than one entrance, 2-way switching is required where the light is switched on at one point and off at another..

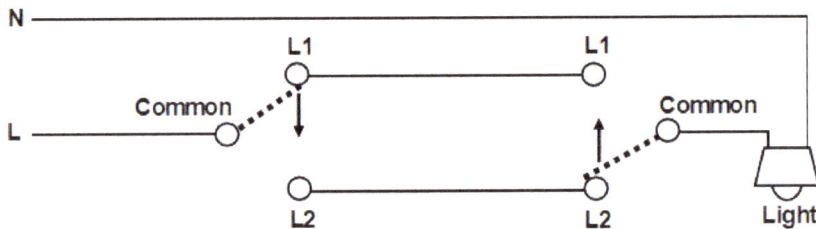

Left is a diagram of a very simple 2-way switch circuit. The dotted lines are the switches and arrows show the movement.

3-Way Switching:

The circuit in the diagram right shows two 2-way switches and an intermediate switch in the centre this could be used where there is a staircase with several floors.

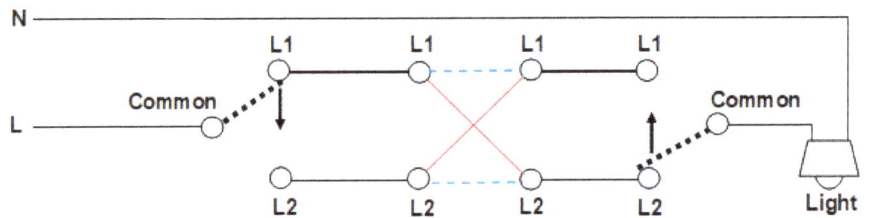

The intermediate switch is a 4-connection switch and its position can be in cross form as per the red solid lines in the diagram **OR** parallel form as per the blue dotted lines. This can be called 3-way light switching or intermediate light switching (in the past it was called corridor light switching). If there are more than three floors then just fit more intermediate switches on them.

Time Delay Switching:

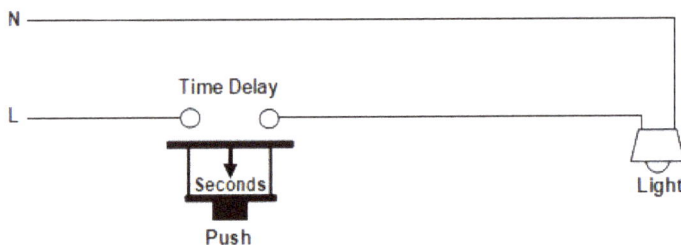

Another type of switch used on stairways is the time delay or dashpot switch. The light is switched on by pushing the plunger on the time delay switch, this will make the contact as in the diagram above and the dashpot will slowly time off.

These can usually be adjusted as per the time. My diagram is a dashpot, but an electronic version can also be obtained. These are usually used on stairways where people keep forgetting to switch off the lights.

Movement Switching:

The switch circuit in the diagram right shows a motion detector where, as soon as you walk into the room it switches the lights on. Not good for some duties such as toilets where if you stop moving and are not in range of the detector you will be in the dark.

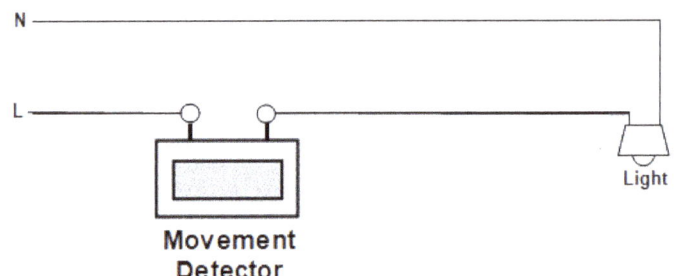

Emergency Lighting (UPS):

Uninterruptible Power Supply (UPS) Systems are used on Chemical Factories for a number of different things and one very important duty is Emergency Lighting. The UPS system allows normal lighting to be used.

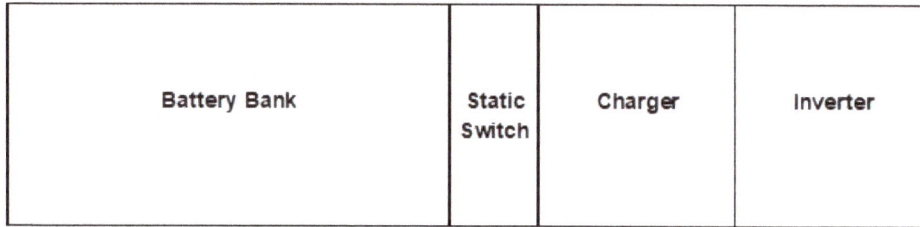

Battery Bank	Static Switch	Charger	Inverter

A small UPS may look similar to the diagram above, where the Batteries are integral to the Inverter and Charger. In other cases, the plant would have a 'battery room' which was full of batteries that may be supplying the Emergency Lighting System.

Usually the Emergency Lighting Inverter feeds normal **240 Volt AC** lighting on the plant. So, in a power outage the inverter would kick in and supply power to the Emergency Lights for some considerable time. On some units the switching can be achieved by an Electronic Static Switch which provides the emergency power when required.

The Lighting Inverter can be tested once a month for, say, 2 hours by manually operating the Static Switch (if fitted). Operators would check the lighting for faulty fluorescent tubes/fittings at the same time. System cut off would ensure that if the lighting was left on to the battery maximum time, then the batteries did not drop below the minimum volts per cell.

Common lamps which **WOULD NOT** be on the emergency lighting system would be, for example, High Pressure Sodium, Low Pressure Sodium and Mercury Arc/Blended. These types of lamps, and others, which require a warm time means there would be a delay before they got to full brightness. Also, if there were to be a series of power outages close together, these lamps have a re-strike time once they have been lit and hot, so the plant could be left in complete darkness for a few minutes which is not desirable if Process or Maintenance Technicians are out there.

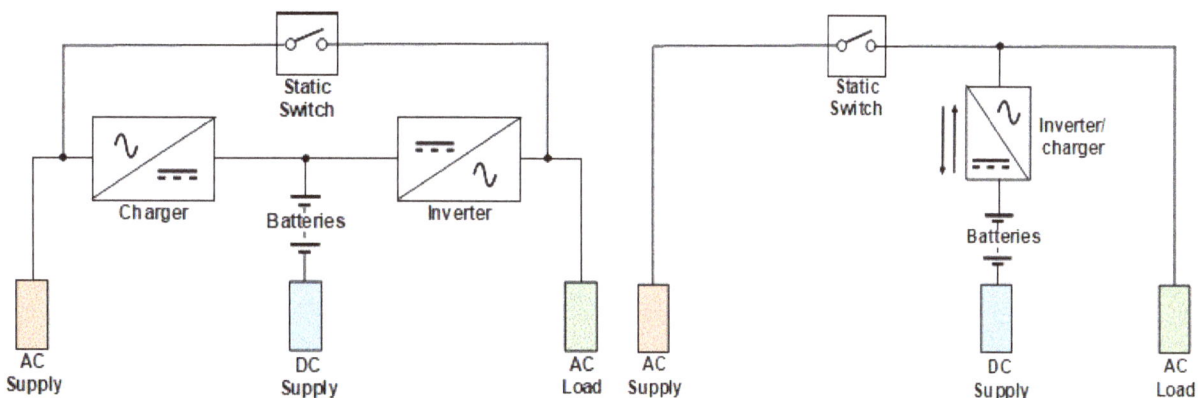

The UPS systems have many different configurations, I have shown two different systems that could be installed:

Above left, has a separate Charger and overall Static Switch. So the Charger charges the Batteries whilst the Inverter is not required and when the Static Switch operates under supply failure and the Inverter takes over.

Above right, shows a Bi-directional Inverter which has an Integral Battery Charger that charges the Batteries whilst everything is normal and then the Inverter will take over in the event of a power loss. There are advantages and disadvantages in both systems and it is always advisory to get Manufacturer's guidance when selecting a system to suit your needs.

Self-contained Lighting:

Self-contained Lighting is lighting with its own Battery unit that will come on in the event of a main power loss. The light that immediately springs to mind is the illuminated 'Exit' signs in the example below, but there are many more examples possibly not so obvious.

Emergency Exit Lights are self-contained units which have a battery back up in the event of a power failure. In the past they would have been a small fluorescent tube, but nowadays they can be obtained as an LED unit. They can maintain battery power for around 3 hours in the event of mains failure. Light level around 1 Lux.

Self-contained Emergency Lighting Unit. These remind me of two car headlights. They are very effective as all that is required is they are plugged into a 13 Amp socket outlet to allow the charging process, and they will automatically light in the event of mains failure. Ideal for public places, and places where there may be a lot of people, such as a back-up for workshops, stores and control rooms. At ground level they are portable allowing the light to be focused anywhere. There are plenty of models on the market, they are not expensive and are ready for use the moment they are removed from their box. Available as LED.

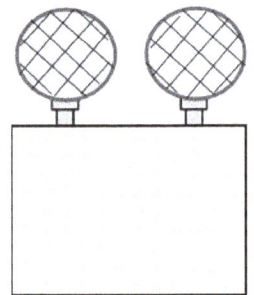

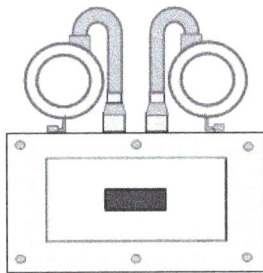

Certified self-contained Emergency Lighting Unit. This model is very similar to the one above except this unit can go into a hazardous area Zone 1 or Zone 2 (Gases & Vapours) & Zone 21 or Zone 22 (Dusts). It can be obtained in Atex certified Exd (Flameproof), Exe (Increased Safety), Extb & Extc (Dust Certified). They are usually 110 or 240 Volts and Ingress Protected to IP66. Can be obtained in LED format. Again there are several makes and models on the market. Can be obtained in Alloy or Aluminium so may have to be careful in Zone 1/Zone 21 Areas.

In the past there would have been a battery power pack on top of a Certified Exd (Flameproof) fluorescent fitting, but things have changed dramatically in the last decade.

Above shows a self-contained fitting that can be installed into a hazardous area Zone 1 or Zone 2 (Gases & Vapours) & Zone 21 or 22 (Dusts) This Fitting would be normally be certified to Atex Dual Certification Exeb (Increased Safety) - mb (Encapsulation) These can probably still be obtained as fluorescent or fluorescent size with LED lamps. Because of the dust certification, their Ingress Protection would be IP66 or IP67. Most self-contained fittings last around 3 hours on battery.

It is possible to obtain self-contained Flameproof Exd lights which can go into dust Zone 21/22 as well as gas Zone 1/2 similar to the diagram right. These lights are LED and are used as emergency escape lighting for areas such as stair wells etc. They give out a 50+ lumen light for around 1 hour.

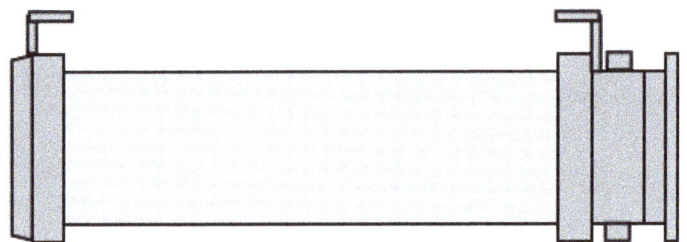

There are two ways that Emergency Lighting can be achieved: a) having a lighting system that is inverter fed or b) having self-contained lighting without a very expensive inverter system. Very similar Atex light fittings can be obtained with motion detection. This depends upon how large the complex and what is best at the time.

Atex Fluorescent Lighting Exe & Exd:

In a hazardous area itself, for example Zone 1 or Zone 2, then the lighting would have to be Atex either Exd Flameproof, Exe Increased Safety or Exn Reduced Risk for gases & vapours or Ext Protection by Enclosure for dust Zone 21 or Zone 22.

Exd Flameproof:

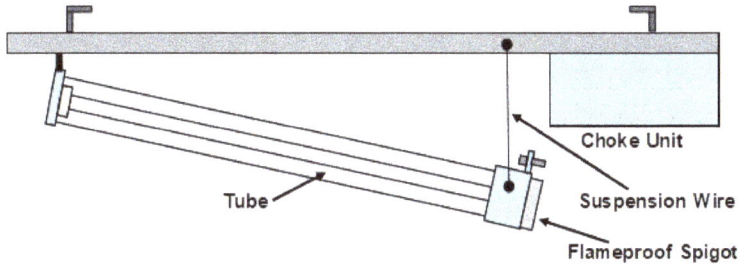

Left is a diagram of a Flameproof Fluorescent Fitting from the 60s. The fitting is open, the tube and glass spigot removed and dropped down on the suspension wire (sometimes a small chain). Remember at this time Flameproof would be BS229 as Atex would not be thought of.

These fittings could easily be mounted the wrong way round with the choke unit end up against a wall in which case you could not remove the tube without undoing the suspension wire. Early versions of this type of fitting were to BS229 (FLP in a Crown) and we had them all over our plants.

Our plants in the 60s were in Divisions not Zones, as we copied the Americans. These days, of course, we are in Zones so this would go into a Zone 1 & Zone 2 Area. Exd Flameproof Fluorescent fittings of today are much more user friendly as far as maintenance is concerned, but there still may be some of the above on very old plants.

Exe Increased Safety: (Atex)

Right is an example of an Exe Increased Safety Fluorescent Fitting. These fittings are not just Exe, but are usually multi-certified for example Exdem suitable for Zone 1 & Zone 2.

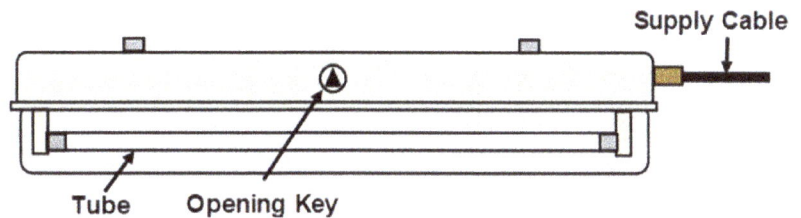

First of all, let us look at the actual fitting. What you are looking at is made out of plastic so this is the Exe Increased Safety. The diffuser is held in position by an interlock system operated by a triangular key in this case. If you do not isolate the fitting, it will isolate itself from where the supply cable enters.

Turning the key does not only undo the interlock and allow the diffuser to open so that the tube(s) can be accessed, but it operates a '**safety switch**' where the supply cable enters the fitting and cuts the power to the rest of the fitting. This safety switch is usually certified Exd Flameproof more because of low volume than Flameproof paths. So the fitting has two different 'protections' or certifications in one enclosure so we have Exde.

Next, we come to the ballast/choke unit. Now this can be 'Encapsulated', in which case it will be certified **Exm Encapsulation**, or it may be Quartz filled, in which case it would be **Exq**. We now have an enclosure with **THREE** 'protections' so here the fitting could be certified Exdemb (Exmb being encapsulated for Zone 1 areas) or Exdeq.

Exd Flameproof. (Atex)

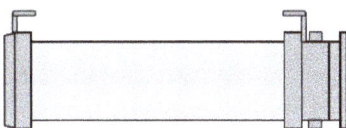

Atex Flameproof Fluorescent Fittings are available, more like the design on the left and as previously described earlier can be obtained as self-contained units. Certification: Exd llC T6.

Hazardous Area Light Fitting Common Designs:

Below are nine lights that are found in Hazardous Area Zones. Firstly, let me say that, there are **NO** lights that are suitable for Zone 0/20 and if you think about it humans would not be in these areas anyway. The nine lights are a mixture of Exd Flameproof, Exe Increased Safety, ExnC Reduced Risk – Enclosed Break & ExnR Reduced Risk – Restricted Breathing. Exd & Exe can be installed in a Zone 1 & 2 Area and Exn can only be installed in a Zone 2 Area.

The above 'Protection' certifications on their own are for Gas & Vapour atmospheres and **ARE NOT** suitable for Dust Zones 21 & 22 as they stand. Manufacturers must design them to Ext also.

You must ensure that the Gas/Dust Group and Temperature Class of the fittings are suitable for the area where you are putting them. Exd Flameproof fittings will be labelled llA, llB or llC, but Exe Increased Safety and Exn Reduced Risk will just have a 'll' meaning **ANY** Gas Group ll. Dust Groups will be in the documentation. Gas & Vapour will be T1-T6, Dust will be surface temperature.

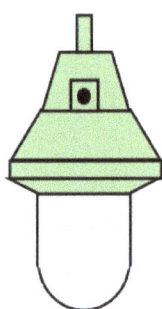

Wellglass: **Exd** Flameproof. These are quite a common fitting and can be installed in Zones 1 & 2. Safety glass will withstand hostile environments. Fitting will be quite heavy and bulky. Can be flanged or spigot enclosure Wellglass.

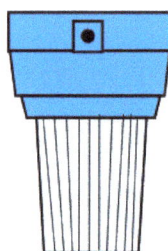

Wellglass: **ExnR** Restricted Breathing. These are quite a common fitting and can be installed in a Zone 2 **ONLY**. Not as heavy as their flameproof counterpart. Usually special conditions apply to their Ingress Protection (IP) and testing.

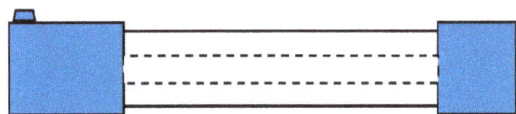

Fluorescent: **Exd** Flameproof. Not so common these days. Can go in Zones 1 & 2. Safety glass. Ballast at one end. Spigot fastening. More expensive than Exe.

Fluorescent: **Exe** Increased Safety. More common these days. Can go in Zones 1 & 2. Plastic diffuser opened with key also operates safety switch.

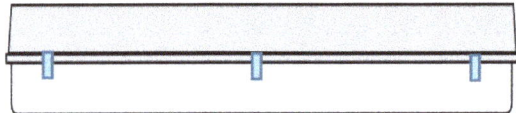

Fluorescent: **Exn** Reduced Risk. More common these days. Can go in a Zones 2 **ONLY**. Plastic diffuser held with clips. No diffuser safety switch.

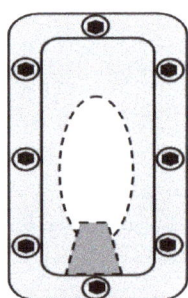

Bulkhead: **Exd** Flameproof. These are not so common these days. Can be installed in Zones 1 & 2. Safety glass will withstand hostile environ-ments. Fitting will be quite heavy and bulky. Usually flanged & special tools required. (i.e., Allen Keys).

Bulkhead: **Exe** increased Safety. These are quite common these days. Can be installed in Zones 1 & 2. The Glass will withstand many hostile environments. IP54 minimum & the installation IP as per manufacturer's instructions. Very similar to ExnR fittings.

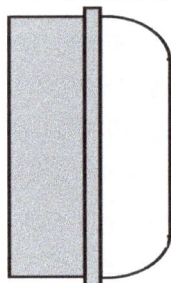

Bulkhead: **ExnC** reduced Risk. These are quite common these days. Can be installed in Zone 2 **ONLY**. The fitting Glass will withstand many of the hostile environments. IP54 minimum & the installation IP as per manufacturer's instructions.

Bulkhead: **ExnR** Restricted Breathing. These are quite a common fitting and can be installed in a Zone 2 **ONLY**. Not as heavy as their Flameproof counterpart. Usually special conditions apply to their Ingress Protection (IP) similar to E

Photocells:

Photocells are used quite a lot on large industrial sites, on both plant lighting and road lighting to switch lights 'on' and 'off' automatically, instead of someone having to go and switch them manually. In the past the older cells would rely on bi-metallic strips for heating and cooling to turn the lights on and off as in the diagram below. There will still be many of this type of photo cell in operation.

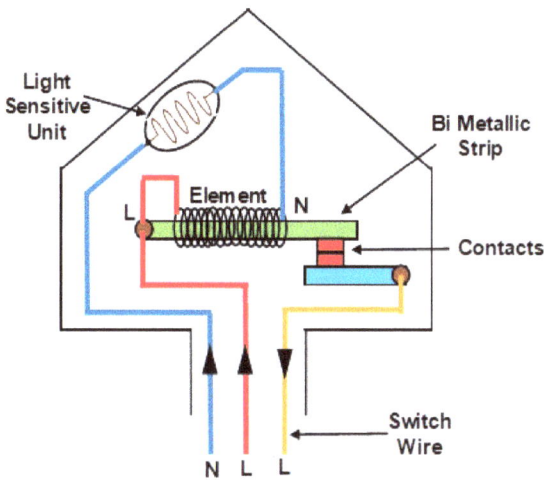

The shape of the cell unit is conical to deter birds from landing on it. The Photo Cell could either be an automatic control of a contactor, which would put power onto a lighting panel, or a control for individual lights such as a street light. As daylight/darkness arrives.

The diagram on the left shows a circuit that could easily be used, as ours were on, say, on plant road lighting. The cell in the diagram has red, blue and yellow wires. The cell works by an element warming a bi-metallic strip and switching the lights **OFF**. No power to the strip switches the lights on, so the unit is failsafe lights on. The light sensitive unit controls the resistance to the neutral to the bi-metallic strip.

Photocells these days are electronic and a lot different from the one that I have drawn in the diagram above. Whichever type they are, they work by photons from the sun falling onto a light sensitive unit, as the one in the diagram, and changing its resistance. Sunlight will make it low resistance and darkness a high resistance.

The cell will work best when it gets what is called 'weathered' and to help this process along, we found that the best adjustment for the cell is to stick small squares of insulation tape over the case directly above the light sensitive unit. This does not sound very technical but it is better than trying adjustment other ways.

The light sensitive unit in the diagram above, when going to high resistance will take away amps from the element on the bi-metallic strip, and the contacts will make switching the lights 'on'. An electronic unit will be slightly different. Here, we are looking at a lower voltage, for a start, and relying on the light sensitive unit to put power 'on' and 'off ' an NPN transistor.

As light and dark falls onto the face of the light sensitive unit the resistance changes to allow current back and forth to operate the circuit and turn the light 'on' and 'off' to the NPN transistor which of course is just an electronic switch, but instead of a toggle, we put a voltage/current.

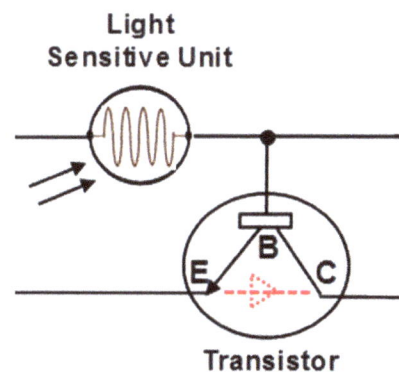

When designing these units and indeed lighting circuits, then the system must be fail safe. So in this case should the Photocell fail then it must fail with **LIGHTS ON** whichever design of cell is used, old or new, and not leave the plant or plant roads in complete darkness.

Because the full circuit relies on light, the location of the photocell is very important. With a road light there is no choice, it fits on top of the lamp post, but if it is switching a full circuit of plant lights then the installation position is very important. It must not be mounted in shadow, e.g., under a buildings eaves, otherwise the lights will come on early before the sun goes down when it is still light. It must not be mounted high in the open, otherwise the sun would have to go fully down over the horizon before the lights came on by which time the lighting level on the plant would be low. It also must not be mounted where other local bright lighting could affect it.

Automatic Lighting Control:

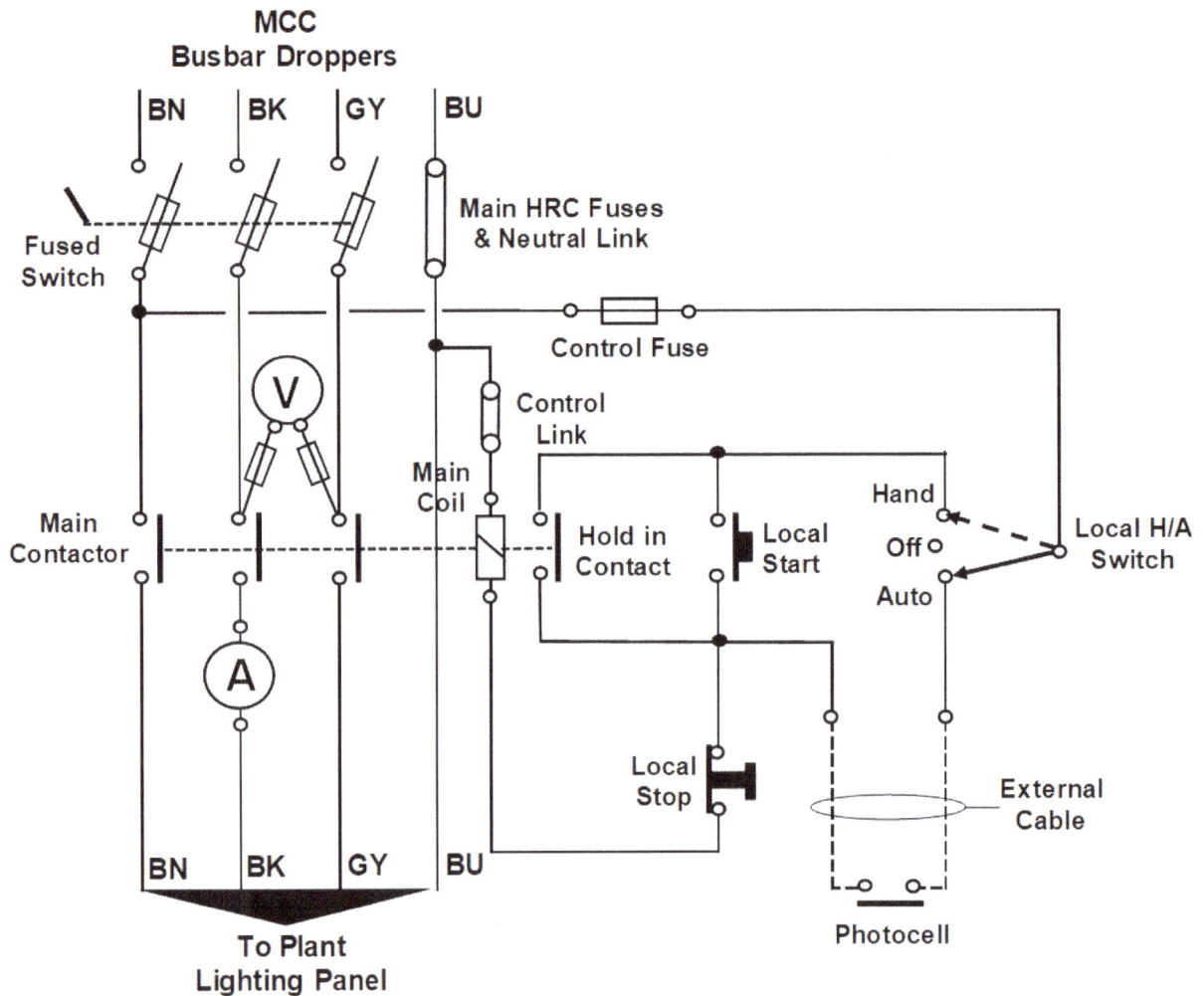

MCC
Busbar Droppers

BN   BK   GY   BU

Main HRC Fuses
& Neutral Link

Fused
Switch

Control Fuse

V

Control
Link

Main
Coil

Main
Contactor

Hold in
Contact

Local
Start

Hand

Off

Auto

Local H/A
Switch

A

Local
Stop

External
Cable

Photocell

BN   BK   GY   BU

To Plant
Lighting Panel

The above diagram is an example of the feed to a plant lighting panel via a 3-phase and neutral contactor. The diagram shows an MCC version, but they can be obtained as independent units. Sometimes, where the contactor is fairly large, an auxiliary relay will be designed into the starter so that there is less current load on the photocell. Ammeters and voltmeters are optional and adhere to Company Policy. Sometimes the ammeter, depending on the load, will be operated by a current transformer (CT). The voltmeter is usually fitted to assist with correct isolation and, in my opinion, a 415V one should be fitted between two phases and not a 240V one between phase and neutral: this is because if the isolator is switched 'off' and the voltmeter goes to zero, it would be correct to assume that all three phases have 'unplugged' and the isolator is square on.

As can be seen in the above diagram, there is a main contactor whose main coil can be energised a) in automatic via a Photocell which is the normal position or b) by hand if the photocell needs to be bypassed and the lights need to be manually lit. If any maintenance is to be carried out on the plant lighting system in a Zoned area, e.g. opening up a light fitting to expose the terminals and cable cores to replace the fitting, the rules are: the basic isolation should be the operation of a double pole switch locked in the 'off' position. A label should be fitted **'Maintenance being carried out on the lighting system'** so that people are aware as to why the lights are isolated. Earlier we discussed that circuits in Zoned areas are on double pole switches, then these would be on the lighting panel.

It is absolutely imperative that no lighting fitting shall be opened up with the circuit switched 'on'. This includes the Photocell. If the system was worked on whilst switched 'on' in automatic and it became dark e.g. nightfall or a storm, the photocell may energise the system.

Light/Lamp Diffusers:

When we talk about lighting diffusers, what do we actually mean? Well, a diffuser is a device used with lighting, especially Fluorescent lighting, to spread, soften or scatter the light and not make it too harsh on the human eye. We have mentioned about how important it is to install warm light, white light etc. in the correct locations and now we have to match that up with the correct diffusers.

Diffusers have to be cleaned at intervals, if you look up at a diffuser that has not been removed for some time you will notice dead flies and dirt etc., and it can end up with the diffuser actually dramatically cutting down the efficiency of the light fitting!

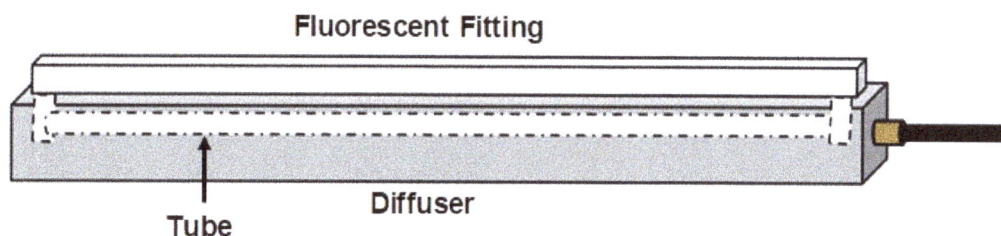

Fluorescent Fitting

Tube   **Diffuser**

Sometimes the diffuser on the Fluorescent fitting is difficult to remove and technicians will develop a knack of removing them. The box diffusers on domestic fittings, such as above, can be made up of hundreds of small prism like shapes to scatter the light or lines of 3D strips which spread the light uniformly.

Although in offices and IT rooms it is advisable to fit lighting with diffusers so that people do not get headaches etc., it must be remembered that diffusers actually cut out at least 20% of the light that the fitting is capable of issuing.

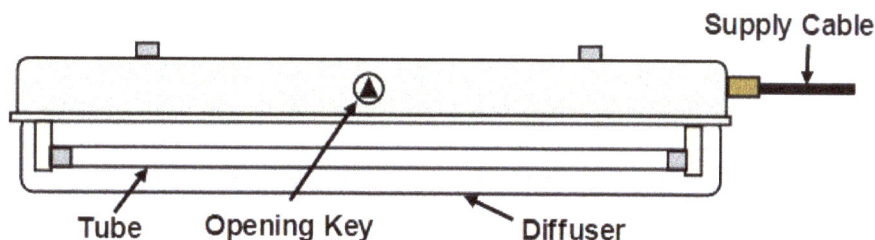

Supply Cable

Tube   **Opening Key**   **Diffuser**

A diffuser on a certified fitting such as Exe, Increased Safety, as mentioned further on in this book is a totally different case. This diffuser can only be opened using a special key, usually triangular, and as well as protecting the tubes and adding to the certification, it also has another function in cutting the power to 7/8ths of the fitting by operating a safety switch when it is opened.

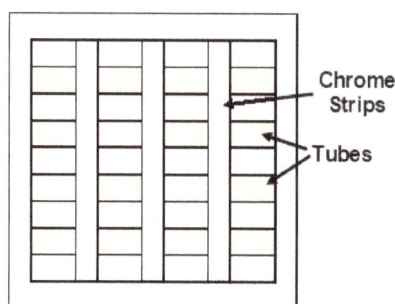

Chrome Strips

Tubes

There must be no harsh lighting in IT rooms or offices with desk monitors, so usually they put this grid type chrome lighting diffuser which gives a very soft light, to prevent any glare on the computer screens. There are many different designs of grid and they work well with LED tubes, which companies are installing to replace fluorescents.

Of course many lamp shades are a type of diffuser and many can be obtained as glass units which will have the prismatic effect of spreading the light. Many shades such as the one on the right are mainly to soften the light and some do this so well that the light efficiency is cut down enormously to the point where the shade is actually blocking the light especially dark coloured ones.

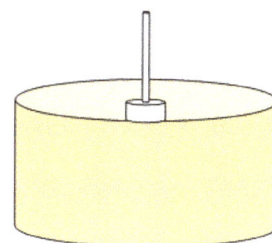

Hazardous Area Lighting Gases & Vapours:

When we talk about lighting, especially for hazardous areas, there are several things that are very important. These are Zones, Gas Groups, Temperature Class & Light Fitting 'Protection'. Let us look at how we would choose our light fittings. Gas & Dust Groups are explained deeper in another section.

Protection: Light fittings come in three types of hazardous area 'protection' namely: Exd (Flameproof), Exe (Increased Safety) & Exn (Reduced Risk). **They can be multi e.g. 'de'**

Zones: In the gases and vapours world there are three Zones to choose from namely: Zone 0, Zone 1 & Zone 2. Let us have a look at the difference:

Zone 0 – Gas or vapour is **present permanently or for very long periods of time.** There are no Zone 0 areas out in the open. We are looking at vapour spaces above the hazardous liquids inside of tanks, vessels and drums. There is **NO** lighting that can be installed in this Zone.

Zone 1 – Gas or vapour is likely to be **present in 'Normal Operation'** which means with the plant running, the gas could be there. So, here we would have to ensure that the light fittings are suitable. We can choose from **Atex Exd (Flameproof) or Atex Exe (Increased Safety)** both can go into the Zone 1 Area.

Zone 2 – Gas or vapour is **unlikely to be present or only for short periods of time** so in 'Normal Operation' which means with the plant running, the gas is unlikely to be present. We would still have to ensure that the light fittings are suitable, just in case, and we can choose from **Atex Exd (Flameproof). Atex Exe (Increased Safety) or Atex Exn (Reduced Risk)** all three can go into the Zone 2 Area.

Gas Groups: There are three main Gas Groups mainly llA, llB & llC. The predominant gases in these groups are Propane - llA, Ethylene - llB and Hydrogen - llC. One of the main Gas Group dividing factors is the amount of 'Ignition Energy' required to ignite the gases. llA gases require around 200µJ (Micro Joules) of energy for ignition (you would almost have to strike the match), llB gases require around 110µJ (Micro Joules) of energy for ignition, but llC gases only require around 20µJ (Micro Joules) of energy only 1/10th of llA gases.

It is imperative therefore to get the right Gas Group. One other group not mentioned above is gas group llB + H**2** which means a llB Gas Group, but also suitable for Hydrogen **NOT** Acetylene or Carbon Di-sulphide. llC Gas Group can go into llA & llB situations provided that the Temperature Classification (T1-T6) and Protection (Exd, Exe & Exn) are suitable. llB Gas Group can go into llA situations provided that the Temperature Classification (T1-T6) and Protection (Exd, Exe & Exn) are suitable. You are not allowed to install llA gas group into llB & llC situations and you **cannot** install llB Gas Group into llC situations. **If there is just a ll then that is ANY gas group ll (A, B or C).**

T. Class	Temp.
T1	450C
T2	400C
T3	300C
T4	135C
T5	100C
T6	85C

Gas or Vapour:	Gas Group:	Ignition Temp.
Carbon Di-sulphide	llC	102C
Cyclohexane	llA	244C
Acetylene	llC	305C
Ethylene	llB	440C
Propane	llA	450C
Hydrogen	llC	560C

Temperature Class (T1-T6): What we are looking at here is ensuring that the **Surface Temperature** of the light fittings **do not go over** the **Ignition Temperature** of the gas or vapour that you are installing it into. The manufacturers state that the **Temperature Class** will not go over the 'T' class stated under **"normal and specified fault conditions"** and they will state any conditions that might occur to take the temperature over the Temperature Class T1-T6. I have included several gases and their approximate Ignition Temperatures above: (Ignition Temperature values vary).

Hazardous Area Lighting Dusts:

When we talk about lighting specially for hazardous areas there are several things which are very important: Zones, Dust Groups, Maximum Surface Temperature & Light Fitting 'Protection'. Let us look at how we would choose our light fittings. Gas & Dust Groups are explained deeper in another section.

Protection:

Dusts: Light fittings come in three main type of hazardous area 'protection' namely: Exta, Extb & Extc. Ext is only a dust protection and does not cover the Gases & Vapours world.

Gases: Exd (Flameproof), Exe (Increased Safety) & Exn (Reduced Risk) are **NOT** dust protections and should be classified into 't' (protection by enclosure) by the manufacturer before they can be used. Manufacturer's guidance is always vital.

Zones:

In the dust world there are three Zones to choose from namely: Zone 20, Zone 21 & Zone 22. Let us have a look at the difference: (Note it is a '2' in front of gas Zones.)

Zone 20 – Dusts present permanently or for very long periods of time. Unlike Gases & Vapours there could be a Zone 20 areas out in the open. There should be **NO** human work in this area without protection. No lighting suitable for Zone 20. Forbidden in work areas.

Zone 21 – Dust is likely to be present in 'Normal Operation' which means with the plant running the dust could be there. Here we would have to ensure that the light fittings are suitable and we can choose from Exta or Extb protection by enclosure.

Atex Exd (Flameproof), Atex Exe (Increased Safety) Atex Exn (Reduced Risk) all of which are **NOT** Dust Zone 21 or 22 protections.

Zone 22 – Dust is unlikely to be present or for short periods of time so in 'Normal Operation' which means with the plant running, the dust is unlikely to be an issue. Here we would still have to ensure that the light fittings are suitable, just in case, and we can choose from Exta, Extb & Extc.

Dust Groups:

There are three main gas groups mainly lllA, lllB & lllC. The predominant gases in these groups are Combustible Flyings - lllA, Non Conductive (e.g. Wood) Dust – lllB and Conductive Dust (e.g. Coal/Carbon) lllC. It is imperative therefore to get the right Dust Group.

Maximum Surface Temperature:

Dust	Ignition Temperature	
	Layer	Cloud
Soot	570°C	810°C
PVC	455°C	700^C
Flour	340°C	490°C
Coffee	220°C	410°C

You must ensure that the surface temperature of the light fittings **will not go over** the ignition temperature of the dust that you are installing it into. The manufacturers state on the equipment the maximum surface temperature, which is always calculated with a 5mm layer of dust on the equipment. Looking at the diagram on the left, the surface temperature of equipment has to be 75C below the dust layer ignition temperature and 2/3rds or 66% below the cloud ignition temperature. Ambient temperature being 40C.

Other specialised equipment must be obtained if the dust layer is over 5mm or the equipment is totally submerged. Dusts also have different ignition temperatures for clouds & layers. (Clouds explode, layers burn)

IP Protection: The equipment must be the correct Ingress Protection (IP) which with dusts will be either IP66 or IP67.

Mining Lamps:

Although we do not have the numbers of miners that we had in the past, we must still look at underground (Subsurface) equipment. When we talk about Atex lighting for Subsurface Industry i.e., Mines, there is a slight difference in the markings to distinguish it from Surface Industry which is explained below. 'Damp' is another name for gas in a mine and one of the main concerns is 'Firedamp' which is a type of Methane and very explosive. The Canary in a cage and Davy lamp warned them of gas present in the past, but detection now is a little more sophisticated. Much of the Electrical equipment that is Exd (Flameproof) or Exia (Intrinsically Safe) Intrinsic Safety was born in the mining industry. Modern Cordless Miner's LED lamps can be obtained and are certified with Lithium battery included with no clumsy belt location.

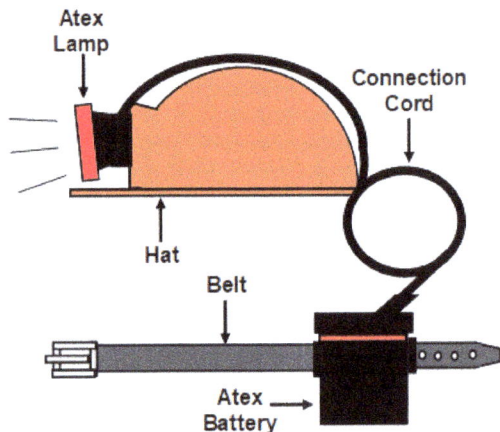

Before we go further, I would like to point out that equipment suitable for 'Subsurface' may not be suitable for 'Surface' and visa-versa. I have chosen a lamp that fits onto a miner's hat but the same marking design would apply to all electrical equipment 'Subsurface'. So, how do we distinguish these two types of different items of equipment? Let us firstly look at the diagram (left).

(I am sure you will appreciate, that in this case the battery, cord and hat lamp must be Atex certified and that miners would require their lighting to be there even if the gas was present.)

Back to the question and diagram. Firstly, all Subsurface electrical equipment, including the Mining Industry will have the following markings:

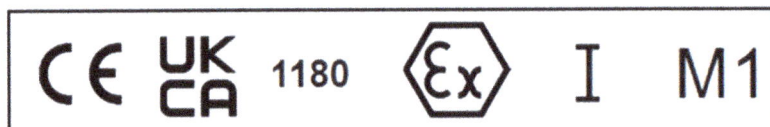

CE – Conformity European which indicates that the equipment is up to European Safety Standards.

UK/CA – Conformity Assessed, now that we have left the European Community because of 'Brexit' if we export Atex equipment to Europe, we have to put this mark onto the equipment to say that the Safety Standards here have been assessed. It may be re-assessed by individual Countries' Health & Safety Bodies whom we export to. Equipment being imported from Europe will just have CE. This CE mark actually stretches beyond Europe.

1180 – Notified Body is the testing organisation that tested the equipment after manufacture to ensure it meets Atex specification. Every Notified body in Europe, and in this case we are included, will have a four-figure number which comes after CE, surface or subsurface industry. I have used 1180 the four-figure number of BASEEFA (**B**ritish **A**pproval **S**ervice for **E**lectrical **E**quipment in **F**lammable **A**tmospheres).

EX in a Hexagon – Is the Atex Mark (**At**mosphere **Ex**plosibles) which in theory should be yellow but manufacturers tend to etch it on the same colour as all the other markings.

Roman 1 after the Atex Mark – This is called Group 1 and Subsurface Industry. This is one of the different markings. If the equipment was for Surface Industry then this would be Group ll and be in the form of a Roman 1 or a Roman ll.

M1 Category – Mining Equipment that must remain energised even if the explosive gas is present e.g. Lighting, breathing air etc. **M2** would be mining equipment that must shutdown if the explosive gas is present e.g. Coal cutting tools etc. Under the **IEC** (**I**nternational **E**lectro-**T**echnical **C**ommission Standards this will change to **Ma & Mb** to match the **EPL** (**E**quipment **P**rotection **L**evels).

ExnR- Restricted Breathing Lighting:

Restricted breathing lighting is an Atex approved method of equipment. The fittings would have the certification markings of ExnR and would be a Category 3 item, which means that it could **ONLY** be installed into a Zone 2 area. I have here focused on lighting, but there are other items of equipment such as instrument panels that are also ExnR. This 'Protection' could soon be re-named to Exec.

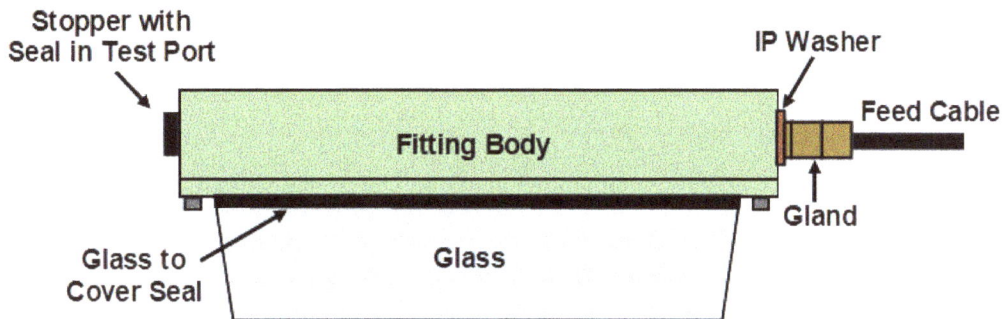

IP washers have to be put onto the glands (whichever IP washers the manufacturers stipulate). There is also a glass-to-cover seal shown in the above diagram, and inside the cover there is another seal to the body. The explanation and reasoning for these seals is below.

What is meant by ExnR Restricted Breathing? Well firstly, the fitting is sealed to a high level, but it is not sealed completely as many people think, so because it can only be installed in Zone 2, a long term concentration of gas outside of the enclosure is unlikely.

When the fitting is lit there is a slight positive pressure inside the enclosure which vents very slowly out to the atmosphere and the reverse happens when it cools down. When the atmosphere is drawn in the IP sealing system limits the entry of the gas, **if any**, to below its lower flammable limit. So in actual fact the fitting breathes through the cover seal in a very restricted way. We do not want it breathing through glands and stoppers so manufacturer's IP seals are put onto these.

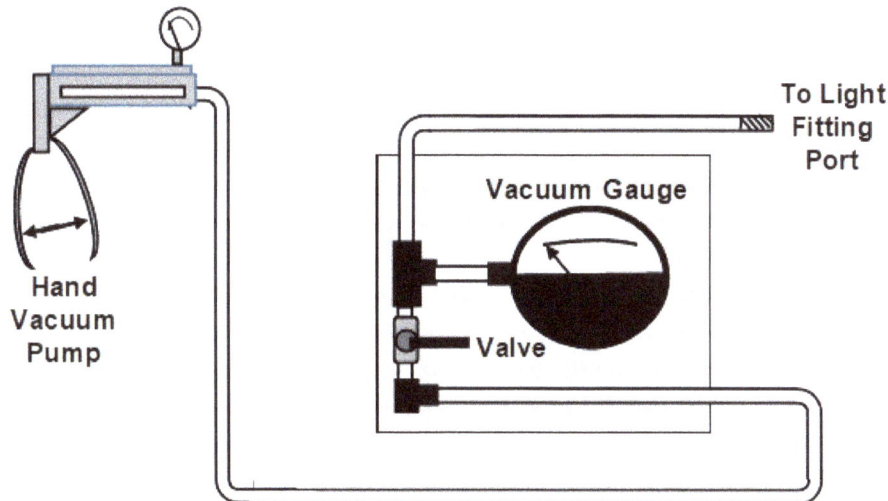

Many enclosures are fitted with a test port sealed with a stopper in the main body to allow the above vacuum test set-up be fitted to the enclosure. This is to test that the restricted breathing property of the enclosure is still satisfactory. This test port must be inspected occasionally to see that it is in good condition.

The above restricted breathing vacuum test set is a hand operated vacuum pump and as mentioned above is connected to the test port, if fitted. This setup will create a vacuum inside of the enclosure to manufacturer's instructions. This may be a routine test to take place after maintenance, but it must be remembered that this vacuum test takes around 10-15 minutes per enclosure.

Finding Hazardous Area Lighting Faults:

I am sure I do not need to emphasise the risks of testing and powering up a circuit when trying to find a lighting fault in a hazardous area, let us say, on a plant on a chemical factory. So, how are we going to go about finding the fault? Safety is uppermost in this situation of both personnel and plant. It is important that a **'Risk Assessment'** be done prior to the work. Company policy must be followed at all times.

So what could cause the lighting fault & why are lamps out?

1 – Broken Cable: This is one of the worst faults. Usually, but not always, if the cable is going to break it will be as it enters a gland and is usually caused by vibration. These days with Steel Wire Armour Cable (SWA) this fault is not so common as in the days of Mineral Insulated Copper Covered Cable (MICC). If the fuse is replaced, there could be sparks and if an Insulation Test is carried out on the cable (500 Volts) there could also be sparks, which are strong enough to ignite a gas. This is not so bad if the area is Zone 2, but a bit more attention should be carried out if the area is Zone 1.

It is absolutely imperative that you complete a Risk Assessment, walk the route of the cable and visually check every fitting, junction box & cable run in the circuit to see if you can see a catastrophic break in the cable or damage before you even think of replacing the fuse.

2 – Lamp Blown: This is a common fault where a lamp blows inside the fitting causing a momentary short circuit and blowing the fuse. Go round every fitting in the circuit and see if there is a lamp laid in the bottom of the well-glass or black emanating from one of the lamps. This is not what I would call a dangerous fault if the fittings are certified Exd Flameproof. **If the fittings are Exe Increased Safety or Exn Reduced Risk then replacing the fuse on a fault could cause sparks to the outside.**

3 – Short Circuit: It is unusual to get a short circuit in a light fitting, junction box or switch, unless the system is very old. If anything, it is caused by a wire coming out of a connector inside of the equipment and touching the case. Maybe an IR check of the system after you have completed the visual check as in '1' above, would be a good idea. Remember, you cannot check for a fault between live and neutral as you have got lamps in the circuit. **Again inside of Exd Flameproof is less of a risk than Exe Increased Safety or Exn Reduced Risk.**

4 – Tired Fuse: Sometimes fuses just blow! On a large complex this may be more common than you think. It is not because there is a fault, but if the fuse is old and has been switching a large bank of lights regularly then the surge reaction on the fuse could make it liable to blow. In this case the fuse can be replaced straight away, but tests 1–3 above would have to be completed first to ensure that it was just the fuse. Also check the fuse size is enough.

5 – Cracked/Broken Well-Glass: Sometimes the glass breaks on the fittings for whatever reason. It is not unusual for the fitting to carry on working on these occasions, but that would be very dangerous on a Zoned area and needs rectifying as soon as possible. If the glass is cracked or broken it is not unusual for water or damp to enter and cause problems.

6 – Seal Damage: If the seal on the light fitting, junction box or switch is damaged then this is an invitation for gas, dust, water or damp to get inside and cause problems. Ensure the seal is completely renewed and not 'make do' even if a new glass or equipment has to be obtained.

7 – Burnt out Equipment: Check for black emanating from the fitting or inside of the glass. Could be the ballast which could be very difficult to detect.

8 – Circuit Isolated: It has been known for technicians to isolate the wrong circuit or isolate a circuit for whatever reason without informing anyone.

9 – Simply Switched off: Ensure that the correct switch is being used to try the circuit. Many technicians have been left red faced trying to find a fault and the circuit is simply switched off.

Double and Single Pole Isolation and Zones:

Firstly let me remind you of the Zones. There are three Zones in the Gas and Vapour world. **Zone 0** where gas or vapour can be there **'Continuously' or for long periods of time,** such as the vapour space above the hazardous liquid in a vessel such as a storage tank, 45-gallon drum, road, rail or sea tanker etc. There will be no lighting, switches or JB.s suitable for this Zone.

Zone 1 is where the gas or vapour is **'Likely'** to be present on, say, a running plant. Finally, **Zone 2** where the gas or vapour is **'Unlikely' to be present or for short periods of time** e.g. leaks which are accidental. The Zones on the plant are worked out on a **'Risk Assessment'** basis by a group of assorted Engineers called an Area Classification Committee.

In the domestic line, switches for lighting are usually single pole with cooker switches and immersion heater switches being double pole: but it does not matter if there is sparking here because there is unlikely to be a hazardous gas present. If you look down into the actual mechanism that you push with your fingers in the dark, you would probably be able to see the spark inside of the switch. This would be of no use in a hazardous area where there may be gas or vapour present, the equipment would have to be Atex certified equipment with Protection (Exd, Exe or Exn), Gas Group (llA, llB or llC) and Temperature Classification (T1-T6) markings.

Dust Zones have a slightly different marking you just put a '2' in front of the Gas Zone number so they are Zone 20, Zone 21 & Zone 22. The rest of the information is the same i.e., Constantly, Likely and Unlikely. Both for Gas and Dust Zones the markings on an Area Classification Drawing are identified on the drawing key.

When working on lighting equipment in a hazardous area, the above is quite different however and you must not be able to see or cause **ANY** sparks to the outside on Atex certified equipment. So the Atex **'Protection'** is very important. Light switches can be obtained with protections Exd Flameproof (Zones 1 & 2), Exe Increased Safety (Zones 1 & 2) or Exn Reduced Risk (Zone 2 only) as can the light fittings. Exd equipment sparks internally, so has to have flameproof paths, be they faces, threads or spigots to stop any internal explosions from propagating to the outside. Exe and Exn equipment must have no internal sparks out in the open and must have an Ingress Protection (IP) of IP54 as a minimum. So, the 'Protection' will cover the equipment zoning, but there can be sparking caused in other ways as below.

THERE ARE NO LIGHT FITTINGS OR SWITCHES SUITABLE FOR ZONE 0 SITUATIONS!

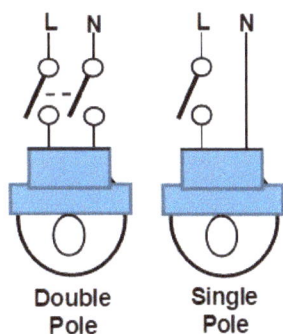

Double Pole   Single Pole

If we take a hazardous area Atex light fitting like the ones in the diagram on the left, and we need to open the connection box for example, disconnecting cables, then we should have in place a double pole isolation locked in that position as minimum. If you just isolate the mains by removing the fuse at the dis-board, the neutral remains connected. it is possible to get accidental neutral to earth sparking, which could at an extreme be around 2-10 volts difference in potential. If the neutral core touches the case of the fitting still connected to earth, this 2-10volt spark is not intrinsically safe and could ignite a gas! Far left is double pole the nearest is single pole.

There is one simple rule to follow: "Operation of a double pole switch padlocked in the OFF position is the minimum isolation requirement when working on the interior of electrical equipment in a hazardous area." So, by locking this unit off we prevent any accidental neutral to earth sparking.

These days, the Atex Certified double pole switches are usually independent units for individual circuits. In the past, there would be one huge lighting panel with numerous double pole switches included on a large busbar section. I am sure that many older plants may still have these panels.

Gas Groups:

The Gas Groups that we are interested in are Group ll Surface Industry whereas in mining (Subsurface) the equipment would be Group l (M1/Ma & M2/Mb). Group ll (Surface Industry) has three sub-divisions, namely llA, llB & llC and these are what we discuss below:

With Gas Groups, we are interested in the Flash Point, where a gas or vapour requires an external means of ignition with enough energy. With the Temperature Classification, you would be more concerned with the Ignition Temperature of the gas or vapour or as it is known: Spontaneous Combustion.

Below I have listed several common materials/gases and their Flash Points, Gas Group & Ignition Temperatures. Remember **flammable liquids don't burn,** vapours burn. So we can see that for the first material Acetone, giving off enough vapours to be ignited with an external ignition at -19 degrees Centigrade. **One** of the main differences between the Gas Groups llA, llB & llC is the different amount of micro-Joules (µJ) for ignition. In the llA Gas Groups for instance would take 10 times the energy from an external ignition to cause combustion than a llC Gas Group.

Material/Gas:	Gas Group:	Degrees Centigrade	
		Flash Point:	Ignition Temp.
Acetone	llA	-19	535
Acetylene	llC	-119	305
Butane	llA	-20	96
Carbon Disulphide	llC	-20	96
Carbon Monoxide	llA	-191	609
Cyclohexane	llA	-18	259
Ethylene	llB	-18	440
Ethanol	llA	12	425
Formic Acid	llA	68	520
Hydrogen	llC	-253	560
Naptha	llA	-6	290
Propane	llA	-104	470
Toluene	llA	6	535
Xylene	llA	30	464

In hazardous areas the slightest spark, unless intrinsically safe, can lead to disaster. As above some gases and vapours do not need many micro-Joules (µJ) for ignition. For example if we were to take a barbeque powered by Propane (C_3H_8) or Butane (C_4H_{10}) these are in what are called the **Gas Group llA** range of gases and would take in the region of **200µJ** of energy to ignite them, so you would almost have to strike the match for combustion. Looking at tables on the Internet, there are many gases in the Gas Group llA range, not quite as many in the llB range, but only three in the llC range - Acetylene, Carbon Disulphide & Hydrogen (H). llC gases only take **20µJ** for ignition, so this is 1/10th the ignitable energy of Propane which makes these gases more dangerous as far as sparks are concerned.

So, Atex light fittings, switches & junction boxes would have to have the correct Gas Group for the worst gas or vapour in the area. Gas group llC equipment being the worst can be installed into Gas Group llA & llB situations and Gas Group llB equipment can be installed into llA situations but you must not go upwards from llA. It is possible to obtain lighting Exd llB+H2 which means the enclosure is for gas group llB, but can be used in a Hydrogen area **NOT** Acetylene or Carbon Disulphide so the gas group could not be put as llC.

Another way that Gas Groups are distinguished is by what is called **Maximum Experimental Safe Gap (MESG).** When manufacturers were carrying out test experiments on their equipment, one of the tests was to find the maximum gap that a gas explosion would not pass through. How they achieved the MESG was to take a box lid and lower the gap between the lid of the box and the box itself.

So looking at the blue box and let us take the gas as Propane, they found that an explosion of **Propane (llA)** would not pass through a gap of 0.92mm.

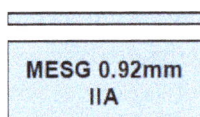

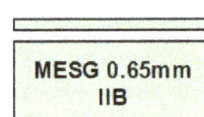

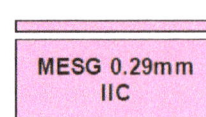

MESG 0.92mm llA   MESG 0.65mm llB   MESG 0.29mm llC

If the explosion will not pass through 0.92mm then it certainly will not pass through a flameproof gap of 0.15mm. Looking now at the green box and let us take the gas as Ethylene, they found that an explosion of **Ethylene (llB)** would not pass through a gap of 0.65mm. So again if the explosion will not pass through 0.65mm it will not pass through a flameproof gap of 0.15mm. Finally the red box and let us take the gas as **Hydrogen (llC),** they found that they got right down to 0.29mm before an explosion of Hydrogen would not pass through the gap so in this case, for safety, they lowered the flameproof gap to 0.1mm which could go down even further depending upon volume.

Dust Groups:

So with **Dust Groups** there is no such thing as a **Flash Point.** Instead there are two different Ignition Temperatures - Layer & Cloud. There is one thing to remember and that is layers burn and clouds explode. Dust Groups are divided into three, the same as Gases & Vapours except that with Dusts the groups are lllA, lllB & lllC. The difference being that **lllA is Combustible Flyings, lllB is Non-Conductive Dust** and **lllC is Conductive Dust. (lllC being the worst).**

Flyings		Non Conductive		Conductive	
Material:	**Group:**	**Material:**	**Group:**	**Material:**	**Group:**
Kapok	lllA	Wood	lllB	Coal	lllC
Jute	lllA	Grain	lllB	Charcoal	lllC
Hemp	lllA	Starch	lllB	Carbon	lllC
Oakum	lllA	Sugar	lllB	Magnesium	lllC

The above tables show several examples of Flyings (lllA), Non-Conductive Dusts (lllB) & Conductive Dusts (lllC). A point about Conductive Dusts is, that you would not want them near to printed circuits or live electrics because of the conductivity. 'Flyings' would be larger than 500 micro-metres (μM) and 'Dusts' would be smaller than 500 micro-metres (μM). Just another point with 'Flyings', you would be able to see them in the air.

There is one protection directly suitable to go into a dust area and that is Ext which is 'Protection by Enclosure i.e., ensuring enclosure seals are suitable and temperature remains low. With Atex this would be derived as ExtD, but now, the IEC Ext is sub divided into Equipment Protection Levels (EPLs) Exta, Extb & Extc.

Zone 20	Zone 21	Zone 22
Exta Only	Exta & Extb	Exta, Extb & Extc
Provided that Dust Group & Temperature Class match.		

So which zones can Ext be installed? As above Exta can be installed into Zone 20, 21 & 22. (provided that the Dust Group above and Temperature Classification match). Extb can be installed into Zone 21 & 22 (providing that the Dust Group and Temperature Classification match). Extc can only be installed into a Zone 22. There are moves that in the future, installation into Zones **MAY** be done on a Risk Assessment basis.

$$\boxed{\textbf{Extb} \quad \textbf{lllB} \quad \textbf{T135}^{\circ}\textbf{C} \quad \textbf{Db}}$$

Switches, light fittings & junction boxes must be installed into the Zones with the correct Dust Group. So, a typical dust marking would be as the example above. The letters mean the following:

Ex – means that the equipment has an Explosion Certificate so it is certified equipment.

tb – 'Protection by Enclosure' and sub-division can be installed in Zone 21 & 22. (IEC)

lllB – this is the Dust Group as explained above.

T135C – This is the maximum surface temperature with a 5mm layer of dust. (IEC)

Db – This is the Equipment Protection Level (EPL) for Zone 21. (IEC)

Hazardous Area Certification Questions:

Below are seven examples of light fitting certification **i.e., Exd llC T6** and several questions. See if you can answer the questions. The answer sheet is on the next page.

1 - Exd llC T2:
1 – What Zone(s) can I put this fitting into?
2 – What Gas Group(s) will this fitting go into?
3 – What is the Maximum Surface Temperature?

2 - Exe ll T5:
1 – What Zone(s) can I put this fitting into?
2 – What Gas Group(s) will this fitting go into?
3 – What is the Maximum Surface Temperature?

3 - Exn ll T4:
1 – What Zone(s) can I put this fitting into?
2 – What Gas Group(s) will this fitting go into?
3 – What is the Maximum Surface Temperature?

4 - Exd llA T3:
1 – What Zone(s) can I put this fitting into?
2 – What Gas Group(s) will this fitting go into?
3 – What is the Maximum Surface Temperature?

5 - Exia llC T6:
1 – What Zone(s) can I put this fitting into?
2 – What Gas Group(s) will this fitting go into?
3 – What is the Maximum Surface Temperature?

6 - Exd llB+H$_2$ T6:
1 – What Zone(s) can I put this fitting into?
2 – What Gas Group(s) will this fitting go into?
3 – What is the Maximum Surface Temperature?

7 - Exta lllB T85ºC:
1 – What Zone(s) can I put this fitting into?
2 – What Dust Group(s) will this fitting go into?
3 – What is the Maximum Surface Temperature?

8 - Extb lllC T100ºC:
1 – What Zone(s) can I put this fitting into?
2 – What Dust Group(s) will this fitting go into?
3 – What is the Maximum Surface Temperature?

Hazardous Area Certification Answers:

Below are the answers to the earlier seven question examples of light fitting certifications I am sure that you got 100%.

1 - Exd llC T2:
(Flameproof)

1 – What Zone(s) can I put this fitting into? **Zones 1 & 2**

2 – What Gas Group(s) will this fitting go into? **llA, llB & llC**

3 – What is the Maximum Surface Temperature? **300C**

2 - Exe ll T5:
(Increased Safety)

1 – What Zone(s) can I put this fitting into? **Zones 1 & 2**

2 – What Gas Group(s) will this fitting go into? **llA, llB & llC**

3 – What is the Maximum Surface Temperature? **100C**

3 - Exn ll T4:
(Reduced Risk)

1 – What Zone(s) can I put this fitting into? **Zone 2 ONLY!**

2 – What Gas Group(s) will this fitting go into? **llA, llB & llC**

3 – What is the Maximum Surface Temperature? **135C**

4 - Exd llA T3:
(Flameproof)

1 – What Zone(s) can I put this fitting into? **Zones 1 & 2**

2 – What Gas Group(s) will this fitting go into? **llA ONLY!**

3 – What is the Maximum Surface Temperature? **200C**

5 - Exia llC T6:
(Intrinsically Safe)

1 – What Zone(s) can I put this fitting into? **Zones 0, 1 & 2**

2 – What Gas Group(s) will this fitting go into? **llA & llB & llC**

3 – What is the Maximum Surface Temperature? **85C**

6 - Exd llB+H$_2$ T6:
(Flameproof)

1 – What Zone(s) can I put this fitting into? **Zones 1 & 2**

2 – What Gas Group(s) will this fitting go into? **llA, llB + Hydrogen**

3 – What is the Maximum Surface Temperature? **85C**

7 - Exta lllB T85ºC:
(Dust Protection)

1 – What Zone(s) can I put this fitting into? **Zones 20, 21 & 22**

2 – What Gas Group(s) will this fitting go into? **lllA & lllB**

3 – What is the Maximum Surface Temperature? **85C**

8 - Extb lllC T100ºC:
(Dust Protection)

1 – What Zone(s) can I put this fitting into? **Zones 21 & 22**

2 – What Dust Group(s) will this fitting go into? **lllA, lllB & lllC**

3 – What is the Maximum Surface Temperature? **100C**

Meaning of Words & Phrases- to do with Light:

Arc Lamp: A lamp that produces light from an arc such as a Carbon arc.

Ballast: A device that limits/stabilises the current. Some people call this device a 'choke' but they are slightly different. A ballast is more resistive and a choke would be more reactive. They both do the same sort of task though. They would be used in some fluorescent fitting, Mercury Vapour, Sodium etc.

Beam Angle: The angle of light deflected by artificial deflector. How far the floodlight or spotlight etc. spreads the light. Spotlight spread is anything from around 5-20 degrees. Floodlights spread the light, depending on the type, from around 20-120 degrees.

Bioluminescence: Light given off by living creatures & organisms.

Black light: The light from a lamp made from Wood's Glass that just emits Ultraviolet light.

Borosilicate: Pyrex glass.

Candela: SI unit of photometry. Latin word for 'Candle'. Sometimes **'Candelas'** are used as another name for Candle Power, but this is only a rough measurement as they are not quite exactly the same! **1 Candle Power = 0.98 Candelas.**

Candescent: The actual term means glowing with heat. Whereas Incandescent means glowing with heat, such as a filament, **and** emitting light as a result.

Candle power: Measure of Luminous Intensity, now obsolete. Many years ago the light **'intensity'** was measured in Candle Power. As you might imagine 1 Candle Power (CP) is the amount of light that would be obtained from 1 candle, 5000 CP is the amount of light from 5000 candles. So this is a measurement of the brightest spot not the overall intensity of an area. **1 Candle Power = 12.57 Lumens.**

Cathodoluminescent: Electrons hitting a luminescent material such as phosphor, the white powder in a fluorescent tube, causing light.

Chemiluminescence: Light given off as a result of a chemical reaction e.g. a light stick.

Choke: A device that limits/stabilises the current by magnetic field. Some people call this device a 'ballast' but they are slightly different. A ballast is more resistive and a choke would be more reactive which affects the power factor. They both do the same sort of task though. They would be used in some fluorescent fitting, Mercury vapour, Sodium etc.

Colour Rendering Index: The amount by which the colour of a surface is lit compared with the standard recommended figures. For instance, if you got white light instead of warm light lamps coloured surfaces may not show their true colour as it could be distorted.

Control Gear: Includes anything that may be used to control a lamp e.g. ballast, choke, starter, ignitor etc. Some lamps, such as a fluorescent tube, require a ballast unit to limit the current and older tubes require a starter.

Dichroic: A type of glass used in LED reflector lights.

Diffused Light: Light that is evenly spread by an artificial diffuser. The light in a computer room, say, may have to be spread evenly otherwise the reflection on the screens may be too harsh for the user.

Diffuser: An artificial device/cover to evenly spread the light. Usually in a computer or photography room. A diffuser softens the light and makes it less harsh. The diffusers come in many designs, some have small prisms that scatter the light and these have to be removed to access the lamp. A diffuser will also cut the light down maybe by 25%.

Discharge Lighting: Lighting that discharges electrons through a gas e.g. Fluorescent Tube, Mercury Blended, Sodium etc. instead of light from a glowing element

Efficacy: The amount of light output by the power (Watts) that are put in.

Ferrite: An Iron Oxide magnetic compound used in the induction coils of induction lighting. Also used in HID lighting to stop the fitting emitting 'Radio Frequency Interference' (RFI) Sometimes used on computer leads and takes the appearance of a black sealed cylinder in the lead containing Ferrite Beads.

Fluorescence: When a substance emits light from radiation that it has absorbed. We could say that it is a form of 'Luminescence'. So we could say that a fluorescent light produces Ultraviolet that is harmful to the human eye, but turns it into fluorescence by coating the tube with phosphor which emits visible light when hit by photons.

Foot Candles: This is actually a measurement of light intensity equivalent to Lumens/Square Foot. So if we took the title virtually it means that if we positioned and lit a candle then this would be the light intensity for an area of 1 foot around the candle. **1 Foot Candle = 10.8 Lux.**

Fresnel Lens: the type of lens used in lighthouses.

Getter: Absorbs unwanted gases in the sodium lamp.

Ghosting: Sometimes when a lamp is switched off the gas is still hot and gives off a faint glow until it cools – this is known as ghosting.

Gobo: A transparent patterned material put in front of a light source to project patterns onto a surface. A movie film may be an example here.

Halide: Halides are a compound of a Halogen (below) atom and another element to form another substance such as a salt.

Halogens: These are reactive non-metals that react with metals to form Halides. There are 5 Halogens:- Astatine, Bromine, Chlorine, Fluorine & Iodine. (Group 7 Periodic Table). They are what is called monovalent elements which readily form negative ions. (Uneven atoms)

High Intensity Discharge (HID): Any gas discharge lamp that works by producing an arc between two electrodes inside of a transparent tube within the bulb filled with inert/noble gas. Sodium, Mercury vapour, Metal Halide etc. instead of a glowing element.

Hydragyrum: Former redundant name for Mercury. This is why Mercury's symbol is Hg.

Illuminance: The amount of light visible to the eye in Lux falling on one spot on a surface.

Incandescent: Light is produced by an extremely hot glowing element.

Inert Gas: Gases that do not react with other elements and are stable. Includes 'Noble' Gases.

Irradiance: Watts per square metre.

Laser: Light Amplification by the Stimulated Emission of Radiation.

LED: Light Emitting Diode.

Lighting Power Density: Amount of power measured in Watts per square foot to light a particular size area.

Light Pollution: Light seen outside of the area that is supposed to be illuminated.

Lumens: Sometimes electrical lighting items are measured in Lumens. So, if we take Candle Power, then all we are measuring here is the amount of light given off by the lighting source, measured by the amount of candles that would give the same light. It has nothing to do with the area size or the distance that the light travels. Lumens only takes into account the area that is illuminated by the light source.

Luminaire: Any whole lamp or fixing. This would include the ballast, diffuser, tubes or bulb, and any brackets etc. that hold the unit in position.

Luminance: The degree and measure of the brightness or luminous intensity of a source of light. The measurement would be in Candelas per square metre. So if you look up at the light you would get the brightness of the lamp onto your retina in Candelas per square metre which with some lamps would be quite a lot. This also applies to light being reflected in a mirror or very shiny surface.

Luminescence: This is the emission of light from a source where there is no heat. An example might be that many watches have the hands giving off luminescence which is a form of radiation without any heat to heat up a filament etc. The hands are said to be luminous and will glow in the dark.

Luminous Efficacy: The amount of illuminance per amount of electrical energy used. Not the same as efficiency. So if you purchase a light bulb and it is 'X' number of Watts exactly how much light are you getting for your Wattage?

Luminous Emittance: Another name for this is, luminous exitance. The Luminous Flux emitted and collecting on an area from a source of light. Measured in Lumens per square metre the similar to luminance.

Luminous Energy Density: This is a measure of the actual amount or intensity emitted by a light source in one direction and is measured in Lumens per Second.

Luminous Exposure: The amount of time a surface is exposed to the visible light in the human spectrum.

Luminous Flux: The power of the light emitted by a light source measured in Lumens.

Luminous Intensity: The measure of the amount of light at a point in one direction.

Lux: Latin word for Light. A measure of illuminance or lumens. (Of light itself.) **1 Lux = 1 Lumen/Square Metre.**

Magnetostriction: Magnetic field forming with the changing current and being absorbed into a fluorescent choke core causing limiting effect with current.

Magnetron: A device for emitting microwave radiation. Dangerous if unguarded.

Mercury Amalgam: Where Mercury is chemically combined with other metals.

Monovalent Element: Readily forms negative ions. (Uneven atoms)

Multifaceted Reflector: The type in lamps called 'downlights' used in kitchens and bathrooms.

Noble Gases: Seven inert 'Noble' gases on the Periodic Table - Helium (He), Neon (Ne), Argon (Ar), Krypton (Kr), Xenon (Xe), Radon (Rn) & Oganesson (Og).

Phosphor: A toxic powder coating that emits light when it meets with a form of energy such as photons in a fluorescent tube. The electrons of the atoms in the phosphor are excited by the electron beam which means they move from a low orbit around the nucleus to a higher one and emit light as they return.

Phosphorescence: This is the action above: The electrons of the atoms in the phosphor are excited by the electron beam which means they move from a low orbit around the nucleus to a higher one and emit light as they return.

Photoluminescence: Light emission from matter after absorbing photons.

Photometry: This is the actual action of measuring the intensity of light visible in the spectrum by the human eye. SI unit 'Candela'

Photon: A particle of light or package of electromagnetic energy with zero mass.

Plasma: The fourth state of very hot matter e.g. Solid, Liquid, Gas & Plasma. It was explained to me in our laboratory, as a sort of gaseous soup of ions or atoms which have free electrons moving around and emitting light.

Polyoxybenzylmethylenglycolanhydride: Bakelite

Quartz Glass: Made from 99% pure Silica (SiO_2)

Radiant Intensity: This is the intensity of electromagnetic radiation. Power per solid angle. So looking back at the lamp from one point on a sphere, shows how much intensity your eye would receive.

Radioluminescence: Light is emitted from a material as a result of radioactive rays hitting it, such as Alpha, Beta or Gamma.

Red Light: Emitted by electromagnetic radiation.

Reflection: This is where light hits a surface and is reflected away as a mirror. Totally different from refraction although people do get them mixed up.

Refraction: This is where light changes direction after passing through a medium for instance a prism. A diffuser may be a good example in this case.

Starter: A device of the past used for causing the HV arc in an older fluorescent tube. Not used any more with modern fluorescent lighting.

Steradian: This is the unit of measurement of the 'solid angle' mentioned in radiant intensity and would take the term square radians.

Stroboscope: A lamp that flashes 'on' and 'off' at a predetermined frequency. Used to measure the speed of rotating objects such as motor shafts. Fluorescent can sometimes be referred to as having a stroboscopic effect.

Thermionic Emission: When a metal, such as a filament, is heated up sufficiently some surface electrons lose their bond and move, hence emitting light in the process.

Troffer: An oblong fluorescent or LED fitting usually located in a lowered false ceiling.

Vitrite: A type of very hard black glass used in lamp/light bulb bases as an insulator for the pins.

Wattage: A lamps energy or equivalent energy consumption. Measurement of light by Wattage. Is really a thing of the past, when incandescent lighting was in full flow. It was easy to see that a 40W, 60W, 100W & 150W were different in their light intensity, but with compact fluorescents and other long-life lamps, this has become more difficult as sometimes lamp manufacturers put confusing light equivalent figures on their packets.

INDEX